Real Analysis and Foundations

Studies in Advanced Mathematics

Volumes in the Series

STEVEN G. KRANTZ
Washington University in St. Louis

Real Analysis and Foundations

CRC PRESS
Boca Raton   Ann Arbor   Boston   London

Library of Congress Cataloging-in-Publication Data

Krantz, Steven G. (Steven George), 1951–
    Real analysis and foundations / Steven G. Krantz.
       p.   cm.
    Includes bibliographical references and index.
    ISBN 0-8493-7156-2
    1. Functions of real variables. 2. Mathematical analysis.
I. Title.
QA331.5.K713   1991
515'.8--dc20   91-26891
                                                    CIP

Direct all inquiries to CRC Press, Inc., 2000 Corporate Blvd., N.W., Boca Raton, Florida, 33431.

© 1991 by CRC Press, Inc.

International Standard Book Number 0-8493-7156-2

Printed in the United States of America   2 3 4 5 6 7 8 9 0

To Stan Philipp, who taught me real analysis.

And to Walter Rudin, who wrote the books from which I learned.

Contents

Preface

Overview

The subject of real analysis, or "advanced calculus," has a central position in undergraduate mathematics education. Yet because of changes in the preparedness of students, and because of their early exposure to calculus (and therefore lack of exposure to certain other topics) in high school, this position has eroded. Students unfamiliar with the value of rigorous, axiomatic mathematics are ill-prepared for a traditional course in mathematical analysis.

Thus there is a need for a book that simultaneously introduces students to rigor, to the *need* for rigor, and to the subject of mathematical analysis. The correct approach, in my view, is not to omit important classical topics like the Weierstrass Approximation Theorem and the Ascoli–Arzela Theorem, but rather to find the simplest and most direct path to each. While mathematics should be written "for the record" in a deductive fashion, proceeding from axioms to special cases, this is *not* how it is learned. Therefore (for example) I *do* treat metric spaces (a topic that has lately been abandoned by many of the current crop of analysis texts). I do so not at first but rather at the end of the book as a method for unifying what has gone before. And I do treat Riemann–Stieltjes integrals, but only after first doing Riemann integrals. I develop real analysis gradually, beginning with treating sentential logic, set theory, and constructing the integers.

The approach taken here results, in a technical sense, in some repetition of ideas. But, again, this is how one learns. Every generation of students comes to the university, and to mathematics, with his or her own viewpoint and background. Thus I have found that the classic texts from which we learned mathematical analysis are often no longer suitable, or appear to be inaccessible, to the present crop of students. It is my hope that my text will be a natural source for modern students to learn mathematical analysis. Unlike other authors, I do not believe that the subject has changed; therefore I have not altered the fundamental content of the course. But the point of view of the audience has changed, and I have written my book accordingly.

The current crop of real analysis texts might lead one to believe that real analysis is simply a rehash of calculus. Nothing could be further from the truth. But many of the texts written thirty years ago are simply too dry and austere for today's audience. My purpose here is to teach today's students the mathematics that I grew to love in a language that speaks to them.

Prerequisites

A student with a standard preparation in lower division mathematics — calculus and differential equations — has adequate preparation for a course based on this text. Many colleges and universities now have a "transitions" course that helps students develop the necessary mathematical maturity for an upper division course such as real analysis. I have taken the extra precaution of providing a mini-transitions course in my Chapters 1 and 2. Here I treat logic, basic set theory, methods of proof, and constructions of the number systems. Along the way, students learn about mathematical induction, equivalence classes, completeness, and many other basic constructs. In the process of reading these chapters, written in a rigorous but inviting fashion, the student should gain both a taste and an appreciation for the use of rigor. While many instructors will want to spend some class time with these two chapters, others will make them assigned reading and begin the course proper with Chapter 3.

How to Build a Course from this Text

Chapters 3 through 7 present a first course in real analysis. I begin with the simplest ideas — sequences of numbers — and proceed to series, topology (on the real line only), limits and continuity of functions, and differentiation of functions. The order of topics is similar to that in traditional books like *Principles of Mathematical Analysis* by Walter Rudin, but the treatment is more gentle. There are many more examples, and much more explanation. I do not shortchange the really interesting topics like compactness and connectedness. The exercise sets provide plenty of drill, in addition to the more traditional "Prove this, Prove that." If it is possible to obtain a simpler presentation by giving up some generality, I always opt for simplicity.

Today many engineers and physicists are required to take a term of real analysis. Chapters 3 through 7 are designed for that purpose. For the more mathematically inclined, this first course serves as an introduction to the more advanced topics treated in the second part of the book.

In Chapter 8 I give a rather traditional treatment of the integral. First the

Riemann integral is covered, then the Riemann–Stieltjes integral. I am careful to establish the latter integral as the natural setting for the integration by parts theorem. I establish explicitly that series are a special case of the Riemann–Stieltjes integral. Functions of bounded variation are treated briefly and their utility in integration theory is explained.

The usual material on sequences and series of functions in Chapter 9 (*including* uniform convergence) is followed by a somewhat novel chapter on "Special Functions." Here I give a rigorous treatment of the elementary transcendental functions as well as an introduction to the gamma function and its application to Stirling's formula. The chapter concludes with an invitation to Fourier series.

I feel strongly, based in part on my own experience as a student, that analysis of several variables is a tough nut the first time around. In particular, college juniors and seniors are not (except perhaps at the very best schools) ready for differential forms. Therefore my treatment of functions of several variables in Chapter 11 is brief, it is only in $\mathbb{R}^3$, and it excludes any reference to differential forms. The main interests of this chapter, from the student's point of view, are (i) that derivatives are best understood using linear algebra and matrices and (ii) that the inverse function theorem and implicit function theorem are exciting new ideas. There are many fine texts that cover differential forms and related material and the instructor who wishes to treat that material in depth should supplement my text with one of those.

Chapter 12 is dessert. For I have waited until now to introduce the language of metric spaces. But now comes the power, for I prove and apply both the Baire category theorem and the Ascoli–Arzela theorem. This is a suitable finish to a year-long course on the elegance and depth of rigorous reasoning.

I would teach my second course in real analysis by covering all of Chapters 8 through 12. Material in Chapters 10 and 12 is easily omitted if time is short.

Audience

This book is intended for college juniors and seniors and some beginning graduate students. It addresses the same niche as the classic books of Apostol, Royden, and Rudin. However, the book is written for today's audience in today's style. All the topics which excited my sense of wonder as a student — the Cantor set, the Weierstrass nowhere-differentiable function, the Weierstrass approximation theorem, the Baire category theorem, the Ascoli–Arzela theorem — are covered. They can be skipped by those teaching a course for which these topics are deemed inappropriate. But they give the subject real texture.

Acknowledgments

It is a pleasure to thank Marco Peloso for reading the entire manuscript of this book and making a number of useful suggestions and corrections. Responsibility for any remaining errors of course resides entirely with me.

Peloso also wrote the solutions manual, which certainly augments the usefulness of the book.

Peter L. Duren, Peter Haskell, Kenneth D. Johnson, and Harold R. Parks served as reviewers of the manuscript that was submitted to CRC Press. Their comments contributed decisively to the clarity and correctness of many passages. I am also grateful to William J. Floyd for a number of helpful remarks.

Russ Hall of CRC Press played an instrumental and propitious role in recruiting me to write for this publishing house. Wayne Yuhasz, Executive Editor of CRC Press, shepherded the project through every step of the production process. Lori Pickert of Archetype, Inc. typeset the book in TEX. All of these good people deserve my sincere thanks for the high quality of the finished book.

— Steven G. Krantz
St. Louis, Missouri

1

Logic and Set Theory

1.1 Introduction

Everyday language is imprecise. Because we are imprecise by *convention*, we can make statements like

All automobiles are not alike.

and feel confident that the listener knows that we actually *mean*

Not all automobiles are alike.

We can also use spurious reasoning like

If it's raining then it's cloudy.

It is not raining.

Therefore there are no clouds.

and not expect to be challenged, because virtually everyone is careless when communicating informally. (Examples of this type will be considered in more detail in Section 1.4.)

Mathematics cannot tolerate this lack of rigor. In order to achieve any depth beyond the most elementary level, we must adhere to strict rules of logic. The purpose of the present chapter is to discuss the foundations of formal reasoning.

In this chapter we will often use numbers to illustrate logical concepts. The number systems we will encounter are

The natural numbers $\mathbb{N} = \{1, 2, 3, \ldots\}$.

The integers $\mathbb{Z} = \{\ldots, -3, -2, -1, 0, 1, 2, 3, \ldots\}$.

The rational numbers $\mathbb{Q} = \{p/q : p \text{ is an integer}, q \text{ is an integer}, q \neq 0\}$.

The real numbers $\mathbb{R}$, consisting of all terminating and nonterminating decimal expansions.

Chapter 2 will be devoted to a thorough and rigorous treatment of number systems. For now we assume that you have seen these number systems before. They are convenient for illustrating the logical principles we are discussing and the fact that we have not constructed them rigorously should lead to no confusion.

1.2 "And" and "Or"

The statement

"A and B"

means that both **A** is true *and* **B** is true. For instance,

George is tall and George is intelligent.

means both that George is tall *and* George is intelligent. If we meet George and he turns out to be short and intelligent, then the statement is false. If he is tall and stupid then the statement is false. Finally, if George is *both* short and stupid then the statement is false. The statement is *true* precisely when both properties — intelligence and tallness — hold. We may summarize these assertions with a *truth table*. We let

A = George is tall.

and

B = George is intelligent.

The expression

$A \wedge B$

will denote the phrase "**A   and   B**." The letters "T" and "F" denote "True" and "False" respectively. Then we have

A	B	$A \wedge B$
T	T	T
T	F	F
F	T	F
F	F	F

Notice that we have listed all possible truth values of **A** and **B** and the corresponding values of the *conjunction* $A \wedge B$.

In a restaurant the menu often contains phrases like

soup or salad

This means that we may select soup *or* select salad, but we may not select both. This use of "or" is called the *exclusive* "or"; it is *not* the meaning of "or" that we use in mathematics and logic. In mathematics we instead say that "**A** or **B**" is true provided that **A** is true or **B** is true or *both* are true. If we let **A** ∨ **B** denote "**A** or **B**" then the truth table is

A	B	A ∨ B
T	T	T
T	F	T
F	T	T
F	F	F

The only way that "**A** or **B**" can be false is if *both* **A** is false and **B** is false. For instance, the statement

Gary is handsome or Gary is rich.

means that Gary is either handsome or rich or both. In particular, he will not be both ugly and poor. Another way of saying this is that if he is poor he will compensate by being handsome; if he is ugly he will compensate by being rich. *But he could be both handsome and rich.*

Example 1.1

The statement

$$x > 5 \text{ and } x < 7$$

is true for the number $x = 11/2$ because this value of x is both greater than 5 *and* less than 7. It is false for $x = 8$ because this x is greater than 5 but not less than 7. It is false for $x = 3$ because this x is less than 7 but not greater than 5. ⬚

Example 1.2

The statement

$$x \text{ is even and } x \text{ is a perfect square}$$

is true for $x = 4$ because both assertions hold. It is false for $x = 2$ because this x, while even, is not a square. It is false for $x = 9$ because this x, while a square, is not even. It is false for $x = 5$ because this x is neither a square nor an even number. ☐

Example 1.3

The statement

$$x > 5 \quad \textbf{or} \quad x \leq 2$$

is true for $x = 1$ since this x is ≤ 2 (even though it is not > 5). It holds for $x = 6$ because this x is > 5 (even though it is not ≤ 2). The statement fails for $x = 3$ since this x is neither > 5 nor ≤ 2. ☐

Example 1.4

The statement

$$x > 5 \quad \textbf{or} \quad x < 7$$

is true for every real x. ☐

Example 1.5

The statement $(\mathbf{A} \vee \mathbf{B}) \wedge \mathbf{B}$ has the following truth table:

A	B	A $\vee$ B	(A $\vee$ B) $\wedge$ B
T	T	T	T
T	F	T	F
F	T	T	T
F	F	F	F

☐

The words "and" and "or" are called *connectives*: their role in sentential logic is to enable us to build up (or connect together) pairs of statements. In the next section we will become acquainted with the other two basic connectives "not" and "if–then."

1.3 "Not" and "If-Then"

The statement "not **A**," written $\sim \mathbf{A}$, is true whenever **A** is false. For example, the statement

Gene is not tall.

is true provided the statement "Gene is tall" is false. The truth table for $\sim$ **A** is as follows:

A	$\sim$ A
T	F
F	T

Although "not" is a simple idea, it can be a powerful tool when used in proofs by contradiction. To prove that a statement **A** is true using proof by contradiction, we instead assume $\sim$ **A**. We then show that this hypothesis leads to a contradiction. Thus $\sim$ **A** must be false; according to the truth table, we see that the only possibility is that **A** is true. We will first encounter proofs by contradiction in Section 1.8.

Greater understanding is obtained by combining connectives:

Example 1.6

Here is the truth table for $\sim (\mathbf{A} \vee \mathbf{B})$:

A	B	A $\vee$ B	$\sim (\mathbf{A} \vee \mathbf{B})$
T	T	T	F
T	F	T	F
F	T	T	F
F	F	F	T

Example 1.7

Now we look at the truth table for $(\sim \mathbf{A}) \wedge (\sim \mathbf{B})$:

A	B	$\sim$ A	$\sim$ B	$(\sim \mathbf{A}) \wedge (\sim \mathbf{B})$
T	T	F	F	F
T	F	F	T	F
F	T	T	F	F
F	F	T	T	T

Notice that the statements $\sim (\mathbf{A} \vee \mathbf{B})$ and $(\sim \mathbf{A}) \wedge (\sim \mathbf{B})$ have the *same truth table*. We call such pairs of statements *logically equivalent*.

The logical equivalence of $\sim (\mathbf{A} \vee \mathbf{B})$ with $(\sim \mathbf{A}) \wedge (\sim \mathbf{B})$ makes good intuitive sense: the statement **A** $\vee$ **B** fails if and only if **A** is false *and* **B** is false. Since in mathematics we cannot rely on our intuition to establish facts, it is important to have the truth table technique for establishing logical equivalence. The exercise set will give you further practice with this notion.

A statement of the form "If **A** then **B**" asserts that whenever **A** is true then **B** is also true. This assertion (or "promise") is tested when **A** is true, because it is then claimed that something else (namely **B**) is true as well. *However*, when **A** is false then the statement "If **A** then **B**" *claims nothing*. Using the symbols **A** $\Rightarrow$ **B** to denote "If **A** then **B**," we obtain the following truth table:

A	B	A $\Rightarrow$ B
T	T	T
T	F	F
F	T	T
F	F	T

Notice that we use here an important principle of Aristotelian logic: every sensible statement is either true or false. There is no "in between" status. Thus when **A** is false then the statement **A** $\Rightarrow$ **B** is not tested. It therefore cannot be false. So it must be true.

Example 1.8

The statement **A** $\Rightarrow$ **B** is logically equivalent with $\sim$ (**A** $\wedge \sim$ **B**). The truth table for the latter is

A	B	$\sim$ B	A $\wedge \sim$ B	$\sim$ (A $\wedge \sim$ B)
T	T	F	F	T
T	F	T	T	F
F	T	F	F	T
F	F	T	F	T

which is the same as the truth table for **A** $\Rightarrow$ **B**.   ⬜

There are in fact infinitely many pairs of logically equivalent statements. But just a few of these equivalences are really important in practice — most others are built up from these few basic ones. The other basic pairs of logically equivalent statements are explored in the exercises.

Example 1.9

The statement

$$\textbf{If } x \textbf{ is negative then } -5 \cdot x \textbf{ is positive.}$$

is true. For if $x < 0$ then $-5 \cdot x$ is indeed > 0; if $x \geq 0$ then the statement is unchallenged. ⬚

Example 1.10
The statement

$$\text{If } \{x > 0 \textbf{ and } x^2 < 0\} \textbf{ then } x \geq 10.$$

is true since the hypothesis "$x > 0$ and $x^2 < 0$" is never true. ⬚

Example 1.11
The statement

$$\text{If } x > 0 \textbf{ then } \{x^2 < 0 \textbf{ or } 2x < 0\}$$

is false since the conclusion "$x^2 < 0$ or $2x < 0$" is false whenever the hypothesis $x > 0$ is true. ⬚

1.4 Contrapositive, Converse, and "Iff"

The statement

$$\textbf{If A then B.} \qquad \text{or} \qquad \textbf{A} \Rightarrow \textbf{B.}$$

is the same as saying

$$\textbf{A suffices for B.}$$

or

$$\textbf{A only if B.}$$

All these forms are encountered in practice, and you should think about them long enough to realize that they say the same thing.

On the other hand,

$$\textbf{If B then A.} \qquad \text{or} \qquad \textbf{B} \Rightarrow \textbf{A.}$$

is the same as saying

$$\textbf{A is necessary for B.}$$

or

$$\textbf{A if B.}$$

We call the statement $\textbf{B} \Rightarrow \textbf{A}$ the *converse* of $\textbf{A} \Rightarrow \textbf{B}$.

Example 1.12

The converse of the statement

> **If x is a healthy horse then x has four legs.**

is the statement

> **If x has four legs then x is a healthy horse.**

Notice that these statements have very different meanings: the first statement is true while the second (its converse) is false. ☐

The statement

> **A if and only if B.**

is a brief way of saying

> **If A then B.**   and   **If B then A**.

We abbreviate **A if and only if B** as **A ⇔ B** or as **A iff B**. Here is a truth table for **A ⇔ B**.

A	B	A ⇒ B	B ⇒ A	A ⇔ B
T	T	T	T	T
T	F	F	T	F
F	T	T	F	F
F	F	T	T	T

Notice that we can say that **A ⇔ B** is true only when both **A ⇒ B** and **B ⇒ A** are true. An examination of the truth table reveals that **A ⇔ B** is true precisely when **A** and **B** are either both true or both false. Thus **A ⇔ B** means precisely that **A** and **B** are logically equivalent. One is true *when and only when* the other is true.

Example 1.13

The statement

$$x > 0 \Leftrightarrow 2x > 0$$

is true. For if $x > 0$ then $2x > 0$; and if $2x > 0$ then $x > 0$. ☐

Example 1.14
The statement

$$x > 0 \Leftrightarrow x^2 > 0$$

is false. For $x > 0 \Rightarrow x^2 > 0$ is certainly true while $x^2 > 0 \Rightarrow x > 0$ is false $((-3)^2 > 0$ but $-3 \not> 0)$. ▯

Example 1.15
The statement

$$\{\sim (\mathbf{A} \vee \mathbf{B})\} \Leftrightarrow \{(\sim \mathbf{A}) \wedge (\sim \mathbf{B})\} \qquad (*)$$

is true because the truth table for $\sim(\mathbf{A} \vee \mathbf{B})$ and that for $(\sim \mathbf{A}) \wedge (\sim \mathbf{B})$ are the same (we noted this fact in the last section). Thus they are logically equivalent: one statement is true precisely when the other is. Another way to see the truth of $(*)$ is to examine the truth table:

A	B	$\sim (\mathbf{A} \vee \mathbf{B})$	$(\sim \mathbf{A}) \wedge (\sim \mathbf{B})$	$\sim (\mathbf{A} \vee \mathbf{B}) \Leftrightarrow \{(\sim \mathbf{A}) \wedge (\sim \mathbf{B})\}$
T	T	F	F	T
T	F	F	F	T
F	T	F	F	T
F	F	T	T	T

▯

Given an implication

$$\mathbf{A} \Rightarrow \mathbf{B},$$

the *contrapositive* statement is defined to be the implication

$$\sim \mathbf{B} \Rightarrow \sim \mathbf{A}.$$

The contrapositive is logically equivalent to the original implication, as we see by examining their truth tables:

A	B	$\mathbf{A} \Rightarrow \mathbf{B}$
T	T	T
T	F	F
F	T	T
F	F	T

and

A	B	$\sim$ A	$\sim$ B	$(\sim$ B$) \Rightarrow (\sim$ A$)$
T	T	F	F	T
T	F	F	T	F
F	T	T	F	T
F	F	T	T	T

Example 1.16
The statement

If it is raining, then it is cloudy.

has, as its contrapositive, the statement

If there are no clouds, then it is not raining.

A moment's thought convinces us that these two statements say the same thing: if there are no clouds, then it couldn't be raining; for the presence of rain implies the presence of clouds.   ⬜

The main point to keep in mind is that, given an implication $\mathbf{A} \Rightarrow \mathbf{B}$, its *converse* $\mathbf{B} \Rightarrow \mathbf{A}$ and its *contrapositive* $(\sim \mathbf{B}) \Rightarrow (\sim \mathbf{A})$ are two different statements. The converse is distinct from, and *logically independent from*, the original statement. The contrapositive is distinct from, but *logically equivalent to*, the original statement.

1.5   Quantifiers

The mathematical statements that we will encounter in practice will use the *connectives* "and," "or," "not," "if–then," and "iff." They will also use *quantifiers*. The two basic quantifiers are "for all" and "there exists."

Example 1.17
Consider the statement

All automobiles have wheels.

This statement makes an assertion about *all* automobiles. It is true, because every automobile does have wheels.

Compare this statement with the next one:

There exists a woman who is blonde.

This statement is of a different nature. It does not claim that all women have blonde hair — merely that there exists *at least one* woman who does. Since that is true, the statement is true. ☐

Example 1.18

Consider the statement

All positive real numbers are integers.

This sentence asserts that something is true for all positive numbers. It is indeed true for *some* positive numbers, such as 1 and 2 and 193. However it is false for at least one positive number, so the entire statement is false.

Here is a more extreme example:

The square of any real number is positive.

This assertion is *almost* true — the only exception is the real number $0 : 0^2 = 0$ is not positive. But it only takes one exception to falsify a "for all" statement. So the assertion is false. ☐

Example 1.19

Look at the statement

There exists a real number that is greater than 5.

In fact there are lots of numbers that are greater than 5; some examples are $7, 42$, and $97/3$. Since there is *at least one* number satisfying the assertion, the assertion is true.

A somewhat different example is the sentence

There exists a real number that satisfies the equation $x^3 - 2x^2 + x - 2 = 0$.

There is in fact only one real number that satisfies the equation, and that is $x = 2$. Yet that information is sufficient to make the statement true. ☐

We often use the symbol $\forall$ to denote "for all" and the symbol $\exists$ to denote "there exists." The assertion

$$\forall x, \ x + 1 < x$$

claims that, for every x, the number $x+1$ is less than x. If we take our universe to be the standard real number system, this statement is false. The assertion

$$\exists x, \; x^2 = x$$

claims that there is a number whose square equals itself. If we take our universe to be the real numbers, then the assertion is satisfied by $x = 0$ and by $x = 1$. Therefore the assertion is true.

Quite often we will encounter $\forall$ and $\exists$ used together. The following examples are typical.

Example 1.20
The statement

$$\forall x \; \exists y, \; y > x$$

claims that for any number x there is a number y that is greater than it. In the realm of the real numbers this is true. In fact $y = x + 1$ will always do the trick.

The statement

$$\exists x \; \forall y, \; y > x$$

has quite a different meaning from the first one. It claims that there is an x that is less than *every* y. This is absurd. For instance, x is *not* less than $y = x - 1$.
▯

Example 1.21
The statement

$$\forall x \; \forall y, \; x^2 + y^2 \geq 0$$

is true in the realm of the real numbers: it claims that the sum of two squares is always greater than or equal to zero.

The statement

$$\exists x \; \exists y, \; x + 2y = 7$$

is true in the realm of the real numbers: it claims that there exist x and y such that $x + 2y = 7$. Certainly the numbers $x = 3, y = 2$ will do the job (although there are many other choices that work as well).   ▯

We conclude by noting that $\forall$ and $\exists$ are closely related. The statements

$$\forall x, \; B(x) \qquad \text{and} \qquad \sim \exists x, \; \sim B(x)$$

are logically equivalent. The first asserts that the statement $B(x)$ is true for all values of x. The second asserts that there exists no value of x for which $B(x)$ fails, which is the same thing.

Likewise, the statements

$$\exists x, B(x) \qquad \text{and} \qquad \sim \forall x, \sim B(x)$$

are logically equivalent. The first asserts that there is some x for which $B(x)$ is true. The second claims that it is not the case that $B(x)$ fails for every x, which is the same thing.

1.6 Set Theory and Venn Diagrams

The two most basic objects in all of mathematics are sets and functions. In this section we discuss the first of these two concepts.

A *set* is a collection of objects. For example, "the set of all blue shirts" and "the set of all lonely whales" are two examples of sets. In mathematics, we often write sets with the following "set-builder" notation:

$$\{x : x + 5 > 0\}.$$

This is read "the set of all x such that $x+5$ is greater than 0. " The universe from which x is chosen (for us this will usually be the real numbers) is understood from context. Notice that the role of x in the set-builder notation is as a *dummy variable*; the set we have just described could also be written as

$$\{s : s + 5 > 0\}$$

or

$$\{\alpha : \alpha + 5 > 0\}.$$

The symbol $\in$ is used to express membership in a set; for example, the statement

$$4 \in \{x : x > 0\}$$

says that 4 is a member of (or *an element of*) the set of all numbers x that are greater than 0. In other words, 4 is a positive number.

If A and B are sets then the statement

$$A \subseteq B$$

is read "A is a subset of B." It means that each element of A is also an element of B (but not *vice versa!*).

Example 1.22

Let

$$A = \{x \in \mathbb{R} : \exists y \text{ such that } x = y^2\}$$

and

$$B = \{t \in \mathbb{R} : t + 3 > -5\}.$$

Then $A \subseteq B$. Why? The set A consists of those numbers that are squares — that is, A is just the nonnegative real numbers. The set B contains all numbers that are greater than -8. Since every nonnegative number (element of A) is also greater than -8 (element of B), it is correct to say that $A \subseteq B$.

However it is not correct to say that $B \subseteq A$, because -2 is an element of B but is not an element of A.   ▯

We write $A = B$ to indicate that both $A \subseteq B$ and $B \subseteq A$. In these circumstances we say that the two sets are equal: every element of A is an element of B and every element of B is an element of A.

We use a slash through the symbols $\in$ or $\subseteq$ to indicate negation:

$$-4 \notin \{x : x \geq -2\}$$

and

$$\{x : x = x^2\} \not\subseteq \{y : y > 1/2\}.$$

It is often useful to combine sets. The set $A \cup B$, called the *union* of A and B, is the set consisting of all objects that are either elements of A *or* elements of B. The set $A \cap B$, called the *intersection* of A and B, is the set consisting of all objects that are elements of *both* A and B.

Example 1.23

Let

$$A = \{x : -4 < x \leq 3\} \ , \quad B = \{x : -1 \leq x < 7\},$$

$$C = \{x : -9 \leq x \leq 12\}.$$

Then

$$A \cup B = \{x : -4 < x < 7\} \ , \quad A \cap B = \{x : -1 \leq x \leq 3\},$$
$$B \cup C = \{x : -9 \leq x \leq 12\} \ , \quad B \cap C = \{x : -1 \leq x < 7\}.$$

Notice that $B \cup C = C$ and $B \cap C = B$ because $B \subseteq C$.   ▯

Example 1.24

Let

$$A = \{\alpha : \alpha \text{ is an integer and } \alpha \geq 9\}$$
$$B = \{\beta \in \mathbb{R} : -4 < \beta \leq 24\},$$
$$C = \{\gamma \in \mathbb{R} : 13 < \gamma \leq 30\}.$$

Then

$$(A \cap B) \cap C = \{x : x \text{ is an integer and } 9 \leq x \leq 24\} \cap C$$
$$= \{t : t \text{ is an integer and } 13 < t \leq 24\}.$$

Also

$$A \cap (B \cup C) = A \cap \{x : -4 < x \leq 30\}$$
$$= \{y : y \text{ is an integer and } 9 \leq x \leq 30\}.$$

Try your hand at calculating $A \cup (B \cup C)$. ▯

The symbol $\emptyset$ is used to denote the set with no elements. For instance,

$$A = \{x \in \mathbb{R} : x^2 < 0\}$$

is a perfectly good set. However there are no real numbers that satisfy the given condition. Thus A is empty, and we write $A = \emptyset$.

Example 1.25

Let

$$A = \{x : x > 8\} \quad \text{and} \quad B = \{x : x^2 < 4\}.$$

Then $A \cup B = \{x : -2 < x < 2 \text{ or } x > 8\}$ while $A \cap B = \emptyset$. ▯

We sometimes use a *Venn diagram* to aid our understanding of set-theoretic relationships. In a Venn diagram, a set is represented as a domain in the plane (for convenience, we use rectangles). The intersection $A \cap B$ of two sets A and

B is the region common to the two domains:

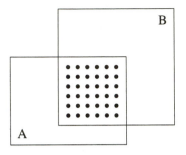

Now let A, B, and C be three sets. The following Venn diagram makes it easy to see that $A \cap (B \cup C) = (A \cap B) \cup (A \cap C)$.

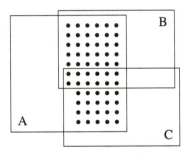

If A and B are sets then $A \setminus B$ denotes those elements that are *in A* but *not in B*. This operation is sometimes called *subtraction of sets* or *set–theoretic difference*.

Example 1.26
Let

$$A = \{x : x > 4\}$$

and

$$B = \{x : x \leq 7\}.$$

Then

$$A \setminus B = \{x : x > 7\}$$

while

$$B \setminus A = \{x : x \leq 4\}.$$

Notice that $A \setminus A = \emptyset$; this fact is true for any set.   ☐

The following Venn diagram illustrates the fact that

$$A \setminus (B \cup C) = (A \setminus B) \cap (A \setminus C).$$

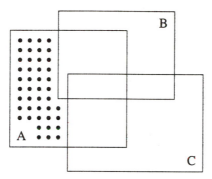

A Venn diagram is not a proper substitute for a rigorous mathematical proof. However it can go a long way toward guiding our intuition.

We conclude this section by mentioning a useful set-theoretic operation and an application. Suppose that we are studying subsets of a fixed set X. If $S \subseteq X$ then we use the notation $^c S$ to denote the set $X \setminus S$. The set $^c S$ is called *the complement of S* (in the set X).

Example 1.27

When we study real analysis, most sets that we consider are subsets of the real line $\mathbb{R}$. If $S = \{x \in \mathbb{R} : 0 \leq x \leq 5\}$ then $^c S = \{x \in \mathbb{R} : x < 0\} \cup \{x \in \mathbb{R} : x > 5\}$. If T is the set of rational numbers then $^c T$ is the set of irrational numbers.

If A, B are sets then it is straightforward to verify that $^c(A \cup B) = {}^c A \cap {}^c B$ and $^c(A \cap B) = {}^c A \cup {}^c B$. More generally, we have

If $\{A_\alpha\}_{\alpha \in A}$ are sets then

$$^c \left(\bigcap_{\alpha \in A} A_\alpha \right) = \bigcup_{\alpha \in A} {}^c A_\alpha$$

and

$$^c \left(\bigcup_{\alpha \in A} A_\alpha \right) = \bigcap_{\alpha \in A} {}^c A_\alpha.$$

The verification of these equalities (known as de Morgan's laws) is left as an exercise. ▯

1.7   Relations and Functions

In more elementary mathematics courses we learn that a "relation" is a rule for associating elements of two sets, and a "function" is a rule that associates to each element of one set a unique element of another set. The trouble with these definitions is that they are imprecise. For example, suppose we define the function $f(x)$ to be identically 1 if there is life as we know it on Mars and to be identically 0 if there is no life as we know it on Mars. Is this a good definition? It certainly is not a very practical one!

More important is the fact that using the word "rule" suggests that functions are given by formulas. Indeed, some functions are, but most are not. Look at any graph in the newspaper — of unemployment, or the value of the Japanese yen, or the Gross National Product. The graphs represent values of these parameters as a function of time. And it is clear that the functions are not given by elementary formulas.

To summarize, we need a notion of function, and of relation, that is precise and flexible and that does not tie us to formulas. We begin with relations and then specialize down to functions.

DEFINITION 1.28   *Let A and B be sets. A **relation** on A and B is a collection of ordered pairs (a, b) such that $a \in A$ and $b \in B$. (Notice that we did not say **the** collection of **all** ordered pairs — that is, a relation consists of some of the ordered pairs, but not necessarily all of them.)*

Example 1.29

Let A be the real numbers and B the integers. The set

$$\mathcal{R} = \{(\pi, 2), (3.4, -2), (\sqrt{2}, 94), (\pi, 50), (2 + \sqrt{17}, -2)\}$$

is a relation on A and B. It associates certain elements of A to certain elements of B. Observe that repetitions are allowed: $\pi \in A$ is associated to both 2 and 50 in B; also $-2 \in B$ is associated to both 3.4 and $2 + \sqrt{17}$ in A.

Now let

$$A = \{3, 17, 28, 42\} \quad \text{and} \quad B = \{10, 20, 30, 40\}.$$

Then

$$\mathcal{R} = \{(3, 10), (3, 20), (3, 30), (3, 40), (17, 20), (17, 30),$$
$$(17, 40), (28, 30), (28, 40)\}$$

is a relation on A and B. In fact $a \in A$ is related to $b \in B$ precisely when $a < b$.   ☐

Example 1.30

Let

$$A = B = \{\text{meter, pound, foot, ton, yard, ounce}\}.$$

Then

$$\mathcal{R} = \{(\text{foot,meter}), (\text{foot,yard}), (\text{meter,yard}), (\text{pound,ton}),$$

$$(\text{pound,ounce}), (\text{ton, ounce}), (\text{meter,foot}), (\text{yard,foot}),$$

$$(\text{yard,meter}), (\text{ton,pound}), (\text{ounce,pound}), (\text{ounce,ton})\}$$

is a relation on A and B. In fact two words are related by $\mathcal{R}$ if and only if they measure the same thing: foot, meter, and yard measure length while pound, ton, and ounce measure weight.

Notice that the pairs in $\mathcal{R}$ are *ordered* pairs: the pair (foot,yard) is different from the pair (yard,foot). ☐

Example 1.31

Let

$$A = \{25, 37, 428, 695\} \qquad \text{and} \qquad B = \{14, 7, 234, 999\}.$$

Then

$$\mathcal{R} = \{(25, 234), (37, 7), (37, 234), (428, 14), (428, 234), (695, 999)\}$$

is a relation on A and B. In fact, two elements are related by $\mathcal{R}$ if and only if they have at least one digit in common. ☐

A function is a special type of relation, as we shall now learn.

DEFINITION 1.32 *Let A and B be sets. A **function** from A to B is a relation $\mathcal{R}$ on A and B such that for each $a \in A$ there is one and only one pair $(a, b) \in \mathcal{R}$. We call A the **domain** of the function and we call B the **range**.*

Example 1.33

Let

$$A = \{1, 2, 3, 4\} \qquad \text{and} \qquad B = \{\alpha, \beta, \gamma, \delta\}.$$

Then

$$\mathcal{R} = \{(1, \gamma), (2, \delta), (3, \gamma), (4, \alpha)\}$$

is a function from A to B. Notice that there is precisely one pair in $\mathcal{R}$ for each element of A. However, notice that repetition of elements of B is allowed. Notice also that there is no apparent "pattern" or "rule" that determines $\mathcal{R}$.

With the same sets A and B consider the relations

$$S = \{(1, \alpha), (2, \beta), (3, \gamma)\}$$

and

$$T = \{(1, \alpha), (2, \beta), (3, \gamma), (4, \delta), (2, \gamma)\}.$$

Then S is not a function because it violates the rule that there be a pair for *each* element of A. Also T is not a function because it violates the rule that there be *just one* pair for each element of A.   ▯

The relations and functions described in the last examples were so simple that you may be wondering what happened to the kinds of functions that we usually look at in mathematics. Now we consider some of those.

Example 1.34
Let $A = \mathbb{R}$ and $B = \mathbb{R}$, where $\mathbb{R}$ denotes the real numbers (to be discussed in detail in Chapter 2). The relation

$$\mathcal{R} = \{(x, \sin x) : x \in A\}$$

is a function. For each $a \in A = \mathbb{R}$ there is one and only one ordered pair with first element a.
     Now let $A = \mathbb{R}$ and $B = \{x \in \mathbb{R} : -2 \leq x \leq 2\}$. Then

$$S = \{(x, \sin x) : x \in A\}$$

is also a function. Technically speaking, it is a different function than $\mathcal{R}$ because it has a different range. However, this distinction often has no practical importance and we shall not mention the difference. It is frequently convenient to write functions like $\mathcal{R}$ or S as

$$\mathcal{R}(x) = \sin x$$

and

$$S(x) = \sin x.$$ ▯

The last example suggests that we distinguish between the set B where a function takes its values and the set of values that the function *actually assumes*.

DEFINITION 1.35   *Let A and B be sets and let f be a function from A to B. Define the **image** of f to be*

$$\text{Image } f = \{b \in B : \exists a \in A \text{ such that } f(a) = b\}.$$

*The set **Image** f is a subset of B.*

Example 1.36

Both the functions $\mathcal{R}$ and $\mathcal{S}$ from the last example have the set $\{x \in \mathbb{R} : -1 \le x \le 1\}$ as image.   ▢

If a function f has domain A and range B and if S is a subset of A then we define

$$f(S) = \{b \in B : b = f(s) \text{ for some } s \in S\}.$$

The set $f(A)$ equals the image of f.

Example 1.37

Let $A = \mathbb{R}$ and $B = \{0, 1\}$. Consider the function

$$f = \{(x, y) : y = 0 \text{ if } x \text{ is rational and } y = 1 \text{ if } x \text{ is irrational}\}.$$

The function f is called the **Dirichlet function** (P. G. Lejeune–Dirichlet, 1805–1859). It is given by a rule, but not by a formula. Notice that $f(\mathbb{Q}) = \{0\}$ and $f(\mathbb{R}) = \{0, 1\}$.   ▢

DEFINITION 1.38   *Let A and B be sets and f a function from A to B.*

*We say that f is **one-to-one** if whenever $(a_1, b) \in f$ and $(a_2, b) \in f$ then $a_1 = a_2$.*

*We say that f is **onto** if whenever $b \in B$ then there exists an $a \in A$ such that $(a, b) \in f$.*

Example 1.39

Let $A = \mathbb{R}$ and $B = \mathbb{R}$. Consider the functions

$$f(x) = 2x + 5, \quad g(x) = \arctan x$$
$$h(x) = \sin x, \quad j(x) = 2x^3 + 9x^2 + 12x + 4.$$

Then f is both one-to-one and onto, g is one-to-one but not onto, j is onto but not one-to-one, and h is neither.

Sketch the graphs of these four functions to convince yourself of these assertions.   ▢

When a function f is both one-to-one and onto then it is called a *bijection* of its domain to its range. In the last example, the function f is a bijection of $\mathbb{R}$ to $\mathbb{R}$.

If f and g are functions, and if the image of g is contained in the domain of f, then we define the *composition* $f \circ g$ to be

$$\{(a, c) : \exists b \text{ such that } g(a) = b \text{ and } f(b) = c\}.$$

This may be written more simply, using the notation introduced in Example 1.34, as

$$f \circ g(a) = f(g(a)) = f(b) = c.$$

Let f have domain A and range B. Assume for simplicity that the image of f is all of B. If there exists a function g with domain B and range A such that

$$f \circ g(b) = b \ \forall b \in B,$$

and

$$g \circ f(a) = a \ \forall a \in A,$$

then g is called the *inverse* of f.

Clearly if the function f is to have an inverse then f must be one-to-one. For if $f(a) = f(a') = b$ then it cannot be that both $g(b) = a$ and $g(b) = a'$. Also f must be onto. For if some $b \in B$ is not in the image of f then it cannot hold that $f \circ g(b) = b$. It turns out that these two conditions are also sufficient for the function f to have an inverse: if f has domain A and range B and if f is both one-to-one and onto then f has an inverse. This matter is explored more thoroughly in the exercises.

Example 1.40

Define a function f, with domain $\mathbb{R}$ and range $\{x \in \mathbb{R} : x \geq 0\}$ by the formula $f(x) = x^2$. Then f is onto but is not one-to-one, hence it cannot have an inverse. This is another way of saying that a positive real number has two square roots, not one.

However the function g, with domain $\{x \in \mathbb{R} : x \geq 0\}$ and range $\{x \in \mathbb{R} : x \geq 0\}$, given by the formula $g(x) = x^2$, does have an inverse. In fact the inverse function is $h(x) = +\sqrt{x}$.

The function $k(x) = x^3$, with domain $\mathbb{R}$ and range $\mathbb{R}$, is both one-to-one and onto. It therefore has an inverse: the function $m(x) = x^{1/3}$ satisfies $k \circ m(x) = x$, and $m \circ k(x) = x$ for all x. □

1.8   Countable and Uncountable Sets

One of the most profound ideas of modern mathematics is Cantor's theory of the infinite (George Cantor, 1845–1918). Cantor's insight was that infinite sets can be compared by size, just as finite sets can. For instance, we think of the number 2 as *less* than the number 3; so a set with two elements is "smaller" than a set with three elements. We would like to have a similar notion of comparison for infinite sets. In this section we will present Cantor's ideas; we will also give precise definitions of the terms "finite" and "infinite."

DEFINITION 1.41 *Let A and B be sets. We say that A and B have the **same cardinality** if there is a function f from A to B that is both one-to-one and onto (that is, f is a bijection from A to B). We write $card(A) = card(B)$.*

Example 1.42
Let $A = \{1, 2, 3, 4, 5\}, B = \{\alpha, \beta, \gamma, \delta, \epsilon\}, C = \{a, b, c, d, e, f\}$. Then A and B have the same cardinality because the function

$$f = \{(1, \alpha), (2, \beta), (3, \gamma), (4, \delta), (5, \epsilon)\}$$

is a bijection of A to B. This function is not the *only* bijection of A to B (can you find another?), but we are only required to produce one.

On the other hand, A and C do not have the same cardinality; neither do B and C. (Find rigorous proofs of these last two statements!) ⬜

Notice that if $card(A) = card(B)$ via a function f_1 and $card(B) = card(C)$ via a function f_2 then $card(A) = card(C)$ via the function $f_2 \circ f_1$.

DEFINITION 1.43 *Let A and B be sets. If there is a one-to-one function from A to B but no bijection between A and B then we will write*

$$card(A) < card(B).$$

This notation is read "A has smaller cardinality than B."
   We use the notation

$$card(A) \leq card(B)$$

to mean that either $card(A) < card(B)$ or $card(A) = card(B)$.

Notice that $card(A) \leq card(B)$ and $card(B) \leq card(C)$ imply that $card(A) \leq card(C)$. Moreover, if $A \subseteq B$ then the inclusion map $i(a) = a$ is a one-to-one function of A into B; therefore $card(A) \leq card(B)$.

The next theorem gives a useful method for comparing the cardinality of two sets.

THEOREM 1.1 SCHROEDER–BERNSTEIN
Let A, B be sets. If there is a one-to-one function $f : A \to B$ and a one-to-one function $g : B \to A$ then A and B have the same cardinality.

PROOF It is convenient to assume that A and B are disjoint; we may do so by replacing A by $\{(a, 0) : a \in A\}$ and B by $\{(b, 1) : b \in B\}$. Let D be the image of f and C be the image of g. Let us define a **chain** to be a sequence of elements of either A or B — that is, a function $\phi : \mathbb{N} \to (A \cup B)$ — such that

$\phi(1) \in B \setminus D$;

if for some j we have $\phi(j) \in B$ then $\phi(j+1) = g(\phi(j))$;

if for some j we have $\phi(j) \in A$ then $\phi(j+1) = f(\phi(j))$.

We see that a chain is a sequence of elements of $A \cup B$ such that the first element is in $B \setminus D$, the second in A, the third in B, and so on. Obviously each element of $B \setminus D$ occurs as the first element of at least one chain.

Define $S = \{a \in A : a \text{ is some term of some chain}\}$. It is helpful to note that

$$S = \{x : x \text{ can be written in the form}$$
$$g(f(g(\cdots g(y) \cdots))) \text{ for some } y \in B \setminus D\}. \qquad (*)$$

We set

$$k(x) = \begin{cases} f(x) & \text{if} \quad x \in A \setminus S \\ g^{-1}(x) & \text{if} \quad x \in S \end{cases}$$

Note that the second half of this definition makes sense because $S \subseteq C$. Then $k : A \to B$. We shall show that in fact k is a bijection.

First notice that f and g^{-1} are one-to-one. This is not quite enough to show that k is one-to-one, but we now reason as follows: If $f(x_1) = g^{-1}(x_2)$ for some $x_1 \in A \setminus S$ and some $x_2 \in S$ then $x_2 = g(f(x_1))$. But, by $(*)$, the fact that $x_2 \in S$ now implies that $x_1 \in S$. That is a contradiction. Hence k is one-to-one.

It remains to show that k is onto. Fix $b \in B$. We seek an $x \in A$ such that $k(x) = b$.

> **Case A:** If $g(b) \in S$ then $k(g(b)) \equiv g^{-1}(g(b)) = b$, hence the x that we seek is $g(b)$.
>
> **Case B:** If $g(b) \notin S$ then we claim that there is an $x \in A$ such that $f(x) = b$. If not then $b \in B \setminus D$. Thus some chain would begin at b. So $g(b)$ would be a term of that chain. We conclude that $g(b) \in S$, and that is a contradiction. So x exists.
>
> Now the x that we just found must lie in $A \setminus S$. For if not then x would be in some chain. Then $f(x)$ and $g(f(x)) = g(b)$ would also lie in that chain. Hence $g(b) \in S$, and that is a contradiction. But $x \in A \setminus S$ tells us that $k(x) = f(x) = b$.

That completes the proof that k is onto. Hence k is a bijection.

The proof of the Schroeder–Bernstein theorem is complete. ∎

Now it is time to look at some specific examples.

Example 1.44
Let E be the set of all even integers and O the set of all odd integers. Then

$$card(E) = card(O).$$

Indeed, the function

$$f(j) = j + 1$$

is a bijection from E to O. ▯

Example 1.45
Let E be the set of even integers. Then

$$card(E) = card(\mathbb{Z}).$$

The function

$$g(j) = j/2$$

is a bijection from E to $\mathbb{Z}$. ▯

This last example is a bit surprising, for it shows that a set can be put in one-to-one correspondence with a proper subset of itself.

Example 1.46
We have

$$card(\mathbb{Z}) = card(\mathbb{N}).$$

We define the function f from $\mathbb{Z}$ to $\mathbb{N}$ as follows:

$f(j) = -(2j + 1)$ if j is negative.
$f(j) = 2j + 2$ if j is positive or zero.

The values that f takes on the negative integers are $1, 3, 5, \ldots$; the values that f takes on the positive integers are $4, 6, 8, \ldots$; and $f(0) = 2$. Thus f is one-to-one and onto. ▯

DEFINITION 1.47 *If a set A has the same cardinality as $\mathbb{N}$ then we say that A is **countable**.*

By putting together the preceding examples, we see that the set of even integers, the set of odd integers, and the set of all integers are countable sets.

Example 1.48

The set of all ordered pairs of positive integers

$$S = \{(j, k) : j, k \in \mathbb{N}\}$$

is countable.

To see this we will use the Schroeder–Bernstein theorem. The function

$$f(j) = (j, 1)$$

is a one-to-one function from $\mathbb{N}$ to S. Also, the function

$$g(j, k) = j \cdot 10^{j+k} + k$$

is a function from S to $\mathbb{N}$. Let n be the number of digits in the number k. Notice that $g(j, k)$ is obtained by writing the digits of j, followed by $j + k - n$ zeroes, then followed by the digits of k. For instance,

$$g(23, 714) = 23 \underbrace{000 \dots 000}_{734} 714,$$

where there are $23 + 714 - 3 = 734$ zeroes between the 3 and the 7. It is clear that g is one-to-one. By the Schroeder–Bernstein theorem, S and $\mathbb{N}$ have the same cardinality; hence S is countable.   ☐

There are other ways to do the last example, and we shall explore them in the exercises.

Since there is a bijection of the set of *all* integers with the set $\mathbb{N}$, it follows from the last example that the set of all pairs of integers (positive *and* negative) is countable.

Notice that the word "countable" is a good descriptive word: if S is a countable set then we can think of S as having a first element (the one corresponding to $1 \in \mathbb{N}$), a second element (the one corresponding to $2 \in \mathbb{N}$), and so forth. Thus we often write $S = \{s_1, s_2, \dots\}$.

DEFINITION 1.49 *A nonempty set S is called **finite** if there is a bijection of S with a set of the form $\{1, 2, \dots, n\}$ for some positive integer n. If no such bijection exists, then the set is called **infinite**.*

An important property of the natural numbers $\mathbb{N}$ is that any subset $S \subseteq \mathbb{N}$ has a least element. This is known as the Well Ordering Principle, and is studied in a course on logic. In the present text we take the properties of the natural numbers as given. We use some of these properties in the next proposition.

PROPOSITION 1.2

If S is a countable set and R is a subset of S then either R is empty or R is finite or R is countable.

PROOF   Assume that R is not empty.

Write $S = \{s_1, s_2, \ldots\}$. Let j_1 be the least positive integer such that $s_{j_1} \in R$. Let j_2 be the least integer following j_1 such that $s_{j_2} \in R$. Continue in this fashion. If the process terminates at the nth step, then R is finite and has n elements.

If the process does not terminate, then we obtain an enumeration of the elements of R :

$$1 \longleftrightarrow s_{j_1}$$

$$2 \longleftrightarrow s_{j_2}$$

$$\ldots$$

etc.

All elements of R are enumerated in this fashion since $j_\ell \geq \ell$. Therefore R is countable.   ∎

A set is called *denumerable* if it is either empty, finite, or countable. In actual practice, mathematicians use the word "countable" to describe sets that are either empty, finite, or countable. In other words, they use the word "countable" interchangeably with the word "denumerable." We shall also indulge in this slight imprecision in this book when no confusion can arise as a result.

The set $\mathbb{Q}$ of all rational numbers consists of all expressions

$$\frac{a}{b}$$

where a and b are integers and $b \neq 0$. Thus $\mathbb{Q}$ can be identified with the set of all pairs (a, b) of integers. After discarding duplicates, such as $\frac{2}{4} = \frac{1}{2}$, and using Example 1.48 and the last Proposition, we find that the set $\mathbb{Q}$ is countable.

THEOREM 1.3

Let S_1, S_2 be countable sets. Set $S = S_1 \cup S_2$. Then S is countable.

PROOF   Let us write

$$S_1 = \left\{ s_1^1, s_2^1, \ldots \right\}$$
$$S_2 = \left\{ s_1^2, s_2^2, \ldots \right\}.$$

If $S_1 \cap S_2 = \emptyset$ then the function

$$s_j^k \mapsto (j, k)$$

is a bijection of S with a subset of $\{(j,k) : j,k \in \mathbb{N}\}$. We proved earlier (Example 1.48) that the set of order pairs of elements of $\mathbb{N}$ is countable. By Proposition 1.2, S is countable as well.

If there exist elements that are common to S_1, S_2 then discard any duplicates. The same argument (use Proposition 1.2) shows that S is countable. ∎

PROPOSITION 1.4

If S and T are each countable sets then so is

$$S \times T = \{(s,t) : s \in S, t \in T\}.$$

PROOF Since S is countable there is a bijection f from S to $\mathbb{N}$. Likewise there is a bijection g from T to $\mathbb{N}$. Therefore the function

$$(f \times g)(s,t) = (f(s), g(t))$$

is a bijection of $S \times T$ with $\mathbb{N} \times \mathbb{N}$, the set of order pairs of positive integers. But we saw in Example 1.48 that the latter is a countable set. Hence so is $S \times T$. ∎

COROLLARY 1.5

If $S_1, S_2, \ldots, S_k$ are each countable sets then so is the set

$$S_1 \times S_2 \times \cdots \times S_k = \{(s_1, \ldots, s_k) : s_1 \in S_1, \ldots, s_k \in S_k\}$$

consisting of all ordered k−tuples $(s_1, s_2, \ldots, s_k)$ with $s_j \in S_j$.

PROOF We may think of $S_1 \times S_2 \times S_3$ as $(S_1 \times S_2) \times S_3$. Since $S_1 \times S_2$ is countable (by the Proposition) and S_3 is countable, then so is $(S_1 \times S_2) \times S_3 = S_1 \times S_2 \times S_3$ countable. Continuing in this fashion, we can see that any finite product of countable sets is also a countable set. ∎

COROLLARY 1.6

The countable union of countable sets is countable.

PROOF Let $A_1, A_2, \ldots$ each be countable sets. If the elements of A_j are enumerated as $\{a_k^j\}$ and if the sets A_j are pairwise disjoint then the correspondence

$$a_k^j \longleftrightarrow (j,k)$$

is one-to-one between the union of the sets A_j and the countable set $\mathbb{N} \times \mathbb{N}$. This proves the result when the sets A_j have no common element. If some of the A_j have elements in common then we discard duplicates in the union and use Proposition 1.2. ∎

PROPOSITION 1.7
The collection $\mathcal{P}$ of all polynomials with integer coefficients is countable.

PROOF Let $\mathcal{P}_k$ be the set of polynomials of degree k with integer coefficients. A polynomial p of degree k has the form

$$p(x) = p_0 + p_1 x + p_2 x^2 + \cdots + p_k x^k.$$

The identification

$$p(x) \longleftrightarrow (p_0, p_1, \ldots, p_k)$$

identifies the elements of $\mathcal{P}_k$ with the $(k+1)$-tuples of integers. By Corollary 1.5, it follows that $\mathcal{P}_k$ is countable. But then Corollary 1.6 implies that

$$\mathcal{P} = \bigcup_{j=0}^{\infty} \mathcal{P}_j$$

is countable. ∎

Georg Cantor's remarkable discovery is that *not all infinite sets are countable*. We next give an example of this phenomenon.

In what follows, a *sequence* on a set S is a function from $\mathbb{N}$ to S. We usually write such a sequence as $s(1), s(2), s(3), \ldots$ or as $s_1, s_2, s_3, \ldots$.

Example 1.50
There exists an infinite set which is not countable (we call such a set **uncountable**). Our example will be the set S of all sequences on the set $\{0, 1\}$. In other words, S is the set of all infinite sequences of 0's and 1's.

To see that S is uncountable, assume the contrary. Then there is a first sequence

$$\mathcal{S}^1 = \left\{ s_j^1 \right\}_{j=1}^{\infty},$$

a second sequence

$$\mathcal{S}^2 = \left\{ s_j^2 \right\}_{j=1}^{\infty},$$

and so forth. This will be a complete enumeration of all the members of S. But now consider the sequence $\mathcal{T} = \{t_j\}_{j=1}^{\infty}$, which we construct as follows:

If $s_1^1 = 0$ then make $t_1 = 1$; if $s_1^1 = 1$ then set $t_1 = 0$.

If $s_2^2 = 0$ then make $t_2 = 1$; if $s_2^2 = 1$ then set $t_2 = 0$.

If $s_3^3 = 0$ then make $t_3 = 1$; if $s_3^3 = 1$ then set $t_3 = 0$.

$$\cdots$$

If $s_j^j = 0$ then make $t_j = 1$; if $s_j^j = 1$ then make $t_j = 0$.

etc.

Now the sequence $\mathcal{T}$ differs from the first sequence $\mathcal{S}^1$ in the first element: $t_1 \neq s_1^1$.

The sequence $\mathcal{T}$ differs from the second sequence $\mathcal{S}^2$ in the second element: $t_2 \neq s_2^2$.

And so on: the sequence $\mathcal{T}$ differs from the jth sequence $\mathcal{S}^j$ in the jth element: $t_j \neq s_j^j$. So the sequence $\mathcal{T}$ is not in the set S. But $\mathcal{T}$ is *supposed* to be in the set S because $\mathcal{T}$ is a sequence of 0's and 1's and all of these are elements of S.

This contradicts our assumption, so S must be uncountable.   ▯

Example 1.51

Consider the set of all decimal representations of numbers — both terminating and nonterminating. Here a terminating decimal is one of the form

$$27.43926$$

while a nonterminating decimal is one of the form

$$3.14159265\ldots.$$

In the case of the nonterminating decimal, no repetition is implied; the decimal simply continues without cease.

Now the set of all those decimals containing only the digits 0 and 1 can be identified in a natural way with the set of sequences containing only 0 and 1 (just put commas between the digits). And we just saw that the set of such sequences is uncountable.

Since the set of all decimal numbers is an even bigger set, it must be uncountable also.

As you may know, the set of all decimals identifies with the set of all real numbers. We find then that the set $\mathbb{R}$ of all real numbers is uncountable. (Contrast this with the situation for the rationals.) In the next chapter we will learn more about how the real number system is constructed using just elementary set theory.   ▯

It is an important result of set theory (due to Cantor) that, given any set S, the set of all subsets of S (called the *power set* of S) has strictly greater cardinality than the set S itself. As a simple example, let $S = \{a, b, c\}$. Then the set of all subsets of S is

$$\Big\{\ \emptyset, \{a\}, \{b\}, \{c\}, \{a,b\}, \{a,c\}, \{b,c\}, \{a,b,c\}\ \Big\}.$$

The set of all subsets has eight elements while the original set has three.

Even more significant is the fact that if S is an infinite set then the set of all its subsets has greater cardinality than S itself. Thus there are infinite sets of arbitrarily large cardinality.

In some of the examples in this section we constructed a bijection between a given set (such as $\mathbb{Z}$) and a proper subset of that set (such as E, the even integers). It follows from the definitions that this is possible only when the sets involved are infinite.

Exercises

1.1 Let the universe be the real number system. Let S = "$x^2 \geq 0$," T = "blue is a primary color," U = "$5 < 3$," and V = "$x > 7$ and $x < 2$." Which of the following statements are true and which are false (use a truth table):
   (a) $S \Longrightarrow T$ (b) $T \Longrightarrow S$
   (c) $S \vee T$ (d) $(\sim S) \wedge U$
   (e) $(\sim U \wedge V)$ (f) $U \vee V$
   (g) $U \vee S$ (h) $\sim (S \Longrightarrow U)$
   (i) $S \Longleftrightarrow V$ (j) $T \Longleftrightarrow U$

1.2 Prove that
   (a) $A \Longrightarrow B$ is logically equivalent to $\sim A \vee B$.
   (b) $A \Longleftrightarrow B$ is logically equivalent to $(\sim (A \wedge (\sim B))) \wedge (\sim (B \wedge (\sim A)))$.
   (c) $A \vee B$ is logically equivalent to $\sim ((\sim A) \wedge (\sim B))$.
   (d) $A \wedge B$ is logically equivalent to $\sim ((\sim A) \vee (\sim B))$.

1.3 The universe is the real numbers. Which of the following statements are true?
   (a) $\forall x \exists y, y < x^2$
   (b) $\exists y \forall x, x^2 + y^2 < -3$
   (c) $\exists x \forall y, y + x^2 > 0$
   (d) $\exists x \forall y, x + y^2 > 0$
   (e) $\forall x \exists y, (x > 0) \Longrightarrow (y > 0 \wedge y^2 = x)$
   (f) $\forall x \exists y, (x > 0) \Longrightarrow (y \leq 0 \wedge y^2 = x)$
   (g) $\forall a \forall b \forall c \exists x, ax^2 + bx + c = 0$

1.4 Write out each of the statements in Exercise 3 using a complete English sentence (no symbols!).

1.5 Let $p(x, y)$ be a statement about the variables x and y. Which of the following pairs of statements are logically equivalent?
   (a) $\forall x \exists y, p(x, y)$ and $\sim \exists x \forall y, \sim p(x, y)$;
   (b) $\forall x \exists y, p(x, y)$ and $\exists y \forall x, p(x, y)$.

1.6 Let the universe be the real number system. Let

$$A = \{x \in \mathbb{R} : x > 0\}, \quad B = \{2, 4, 8, 16, 32\},$$

$$C = \{2, 4, 6, 8, 10, 12, 14\},$$

$$D = \{x : -3 < x < 9\}, \quad E = \{x : x \leq 1\}.$$

Calculate the six sets

$$B \cap C, \quad B \cup C, \quad A \cap (D \cup E),$$

$$A \cup (B \cap C), \quad (A \cap C) \cup (B \cap D),$$

$$A \cap (B \cap (C \cap (D \cap E))).$$

1.7 Which of the following sets are countable and which are not (provide detailed justification for your answers):

(a) the set of irrational numbers *no*

(b) the set of terminating decimals

(c) the set of real numbers between 0.357 and 0.358

(d) $\mathbb{Q} \times \mathbb{Q}$ *countable*

(e) the set of numbers obtained from $\sqrt{2}$ and $\sqrt{3}$ by finitely many arithmetic operations $(+, -, \times, \div)$

(f) $\mathbb{N} \times \mathbb{Z}$

(g) $\mathbb{R} \times \mathbb{Z}$ *R=uncountable*

1.8 Is the intersection of two countable sets countable? How about their union?

1.9 Is the intersection of two uncountable sets uncountable? How about their union?

1.10 Let A, B, C, D be sets. Sketch Venn diagrams to illustrate each of the following:

(a) $A \cup B$ (b) $A \cup (B \cap C)$

(c) $C \setminus (B \cup C)$ (d) $C \setminus (B \cap A)$

(e) $C \cap (B \cap A)$ (f) $A \cup (B \cup C)$

1.11 Let A, B, C be sets. Prove each of the following statements:

$$C \setminus (A \cup B) = (C \setminus A) \cap (C \setminus B).$$

$$C \setminus (A \cap B) = (C \setminus A) \cup (C \setminus B).$$

(Hint: a Venn diagram is not a proof.)

1.12 Consider the set $S = \mathbb{N} \times \mathbb{N}$ of all ordered pairs of positive integers. Write the elements of S in an array as follows:

$$
\begin{array}{cccccc}
(1,1) & (1,2) & (1,3) & (1,4) & (1,5) & \cdots \\
(2,1) & (2,2) & (2,3) & (2,4) & (2,5) & \cdots \\
(3,1) & (3,2) & (3,3) & (3,4) & (3,5) & \cdots \\
(4,1) & (4,2) & (4,3) & (4,4) & (4,5) & \cdots \\
(5,1) & (5,2) & (5,3) & (5,4) & (5,5) & \cdots \\
\end{array}
$$

$$\cdots$$

Enumerate the pairs by counting along *diagonals* that extend from the lower left to the upper right. This gives an alternate way to prove that $\mathbb{N} \times \mathbb{N}$ is countable.

1.13 Prove that if a function f, with domain A and range B, is both one-to-one and onto then f has an inverse function g.

1.14 Consider the statement

$$\text{If } x > 2 \text{ then } x^2 > 6.$$

Explain why the statement is true for $x = 3$. Explain why the statement is true for $x = 1$. Explain why the statement is true for $x = -4$. Explain why the statement is false for $x = 2.1$. Do *not* use truth tables!

1.15 If $A_1, A_2, \ldots$ are sets then define

$$\prod_{j=1}^{\infty} A_j$$

to be the collection of all functions from the natural numbers $\mathbb{N}$ into $\cup A_j$ such that $f(j) \in A_j$. What can you say about the cardinality of the set

$$\prod_{j=1}^{\infty} A_j$$

when each A_j has the cardinality of $\mathbb{Z}$? What about when each of the A_j has the cardinality of $\mathbb{R}$?

1.16 Consider the set S of all real numbers obtained by taking rational powers of rational numbers. Is this set countable or uncountable?

1.17 A subset S of the plane is called *convex* if whenever $a, b \in S$ then the line segment connecting a to b lies in S. What is the cardinality of the collection of convex sets in the plane?

1.18 Give an explicit example of a set that has cardinality *greater* than the cardinality of the set of all real numbers and prove that the cardinality is greater.

1.19 Prove that it is impossible for a finite set to be put in one-to-one correspondence with a proper subset of itself.

1.20 Let S be an infinite set. Prove that there is a subset $T \subseteq S$ such that T is countable.

1.21 Prove that it is always possible to put an infinite set in one-to-one correspondence with a proper subset of itself. [Hint: Consider the natural numbers first. Then use Exercise 20 to treat the general case.]

1.22 What is the cardinality of $\mathbb{R} \times \mathbb{N}$?

1.23 What is the cardinality of $\mathbb{R} \times \mathbb{R}$?

1.24 Consider the statement

$$A \Longrightarrow B \Longrightarrow C.$$

Write a truth table for this statement. Can you do this without inserting parentheses? Does your answer depend on where you insert the parentheses? Discuss the possibilities.

1.25 Repeat Exercise 24 with "$\Rightarrow$" replaced by $\wedge$.

1.26 Repeat Exercise 24 with "$\Rightarrow$" replaced by $\vee$.

1.27 Let S be the set of all *finite* sequences of 0's and 1's. Is this set countable or uncountable?

1.28 If A is uncountable and B is uncountable then what can you say about $\{f : f$ is a function from A to $B\}$?

2

Number Systems

2.1 The Natural Numbers

Mathematics deals with a variety of number systems. The simplest number system in real analysis is $\mathbb{N}$, the *natural numbers*. As we have already noted, this is just the set of positive integers $\{1, 2, 3, \ldots\}$. In a rigorous course of logic, the set $\mathbb{N}$ is constructed from the axioms of set theory. However, in this book we shall assume that you are familiar with the positive integers and their elementary properties.

The principal properties of $\mathbb{N}$ are as follows:

1. 1 is a natural number.

2. If x is a natural number then there is another natural number $\hat{x}$ which is called the *successor* of x.

3. $1 \neq \hat{x}$ for every natural number x.

4. If $\hat{x} = \hat{y}$ then $x = y$.

5. **(Principle of Induction)** If Q is a property and if

   (a) 1 has the property Q;

   (b) whenever a natural number x has the property Q it follows that $\hat{x}$ also has the property Q;

   then all natural numbers have the property Q.

These rules, or *axioms*, are known as the Peano Axioms for the natural numbers (named after Giuseppe Peano (1858–1932) who developed them). We take it for granted that the usual set of positive integers satisfies these rules. Certainly 1 is in that set. Each positive integers has a "successor" — after 1 comes 2 and after 2 comes 3 and so forth. The number 1 is not the successor of any other positive integer. Two positive integers with the same successor must be the same. The last axiom is more subtle but makes good sense: if some property

$Q(n)$ holds for $n = 1$ and if whenever it holds for n then it also holds for $n + 1$, then we may conclude that Q holds for all positive integers.

We will spend the rest of this section exploring Axiom (5), the Principle of Induction.

Example 2.1

Let us prove that for each positive integer n it holds that

$$1 + 2 + \cdots + n = \frac{n \cdot (n + 1)}{2}.$$

We denote this equation by $Q(n)$ and follow the scheme of the Principle of Induction.

First, $Q(1)$ is true since then both the left and the right side of the equation equal 1. Now assume that $Q(n)$ is true for some natural number n. Our job is to show that it follows that $Q(n + 1)$ is true.

Since $Q(n)$ is true, we know that

$$1 + 2 + \cdots + n = \frac{n \cdot (n + 1)}{2}.$$

Let us add the quantity $n + 1$ to both sides. Thus

$$1 + 2 + \cdots + n + (n + 1) = \frac{n \cdot (n + 1)}{2} + (n + 1).$$

The right side of this new equality simplifies and we obtain

$$1 + 2 + \cdots + (n + 1) = \frac{(n + 1) \cdot ((n + 1) + 1)}{2}.$$

But this is just $Q(n + 1)$ or $Q(\hat{n})$! *We have assumed $Q(n)$ and have used it to prove $Q(\hat{n})$*, just as the Principle of Induction requires.

Thus we may conclude that property Q holds for all positive integers, as desired. ☐

The formula which we derived in the first example was probably known to the ancient Greeks. However, a celebrated anecdote credits Carl Friedrich Gauss (1777–1855) with discovering the formula when he was nine years old. Gauss went on to become (along with Isaac Newton and Archimedes) one of the three greatest mathematicians of all time.

The formula from Example 2.1 gives a neat way to add up the integers from 1 to n, for any n, without doing any work. Any time that we discover a new mathematical fact, there are generally several others hidden within it. The next example illustrates this point.

Example 2.2

The sum of the first m positive even integers is $m \cdot (m + 1)$. To see this note that the sum in question is

$$2 + 4 + 6 + \cdots + 2m = 2(1 + 2 + 3 + \cdots + m).$$

But, by the first example, the sum in parentheses on the right is equal to $m \cdot (m + 1)/2$. It follows that

$$2 + 4 + 6 + \cdots + 2m = 2 \cdot \frac{m \cdot (m + 1)}{2} = m \cdot (m + 1). \qquad \Box$$

The second example could also be performed by induction (without using the result of the first example). This method is explored in the exercises.

Example 2.3

Now we will use induction incorrectly to prove a statement that is completely preposterous:

All horses are the same color.

There are finitely many horses in existence, so it is convenient for us to prove the slightly more technical statement

Any collection of k horses consists of horses that are all the same color.

Our statement $Q(k)$ is this last displayed statement.

Now $Q(1)$ is true: *one horse is the same color.* (Note: this is not a joke, and the error has not occurred yet.) Suppose that $Q(k)$ is true: we assume that any collection of k horses has the same color. Now consider a collection of $\hat{k} = k + 1$ horses. Remove one horse from that collection. By our hypothesis, the remaining k horses have the same color.

Now replace the horse that we removed and remove a different horse. Again, the remaining k horses have the same color.

We keep repeating this process: remove each of the $k + 1$ horses one by one and conclude that the remaining k horses have the same color. Therefore every horse in the collection is the same color as every other. So all $k + 1$ horses have the same color. The statement $Q(k + 1)$ is thus proved (assuming the truth of $Q(k)$) and the induction is complete.

Where is our error? It is nothing deep — just an oversight. The argument we have given is wrong when $\hat{k} = k + 1 = 2$. For remove one horse from a set of two and the remaining (*one*) horse is the same color. Now replace the removed horse and remove the other horse. The remaining (*one*) horse is the same color. *So what?* We cannot conclude that the two horses are colored the same. Thus the induction breaks down at the outset; the reasoning is incorrect. $\Box$

PROPOSITION 2.1   THE BINOMIAL THEOREM

Let a and b be real numbers and n a natural number. Then

$$(a+b)^n = a^n + \frac{n}{1}a^{n-1}b + \frac{n(n-1)}{2\cdot 1}a^{n-2}b^2$$

$$+ \frac{n(n-1)(n-2)}{3\cdot 2\cdot 1}a^{n-3}b^3$$

$$+ \ldots + \frac{n(n-1)\cdots 2}{(n-1)(n-2)\cdots 2\cdot 1}ab^{n-1} + b^n.$$

PROOF   The case $n=1$ being obvious, proceed by induction.   ∎

REMARK 2.2   The expression

$$\frac{n(n-1)\cdots(n-k+1)}{k(k-1)\cdots 1}$$

is often called the kth *binomial coefficient* and is denoted by the symbol

$$\binom{n}{k}.$$

Using the notation $m! = m\cdot(m-1)\cdot(m-2)\cdots 2\cdot 1$, for m a natural number, we may write the kth binomial coefficient as

$$\binom{n}{k} = \frac{n!}{(n-k)!\cdot k!}.$$ ∎

2.2   Equivalence Relations and Equivalence Classes

Let S be a set and let $\mathcal{R}$ be a relation on S and S. We call $\mathcal{R}$ an *equivalence relation* on S if $\mathcal{R}$ has the following three properties:

(Reflexivity) If $s \in S$ then $(s,s) \in \mathcal{R}$.

(Symmetry) If $(s,t) \in \mathcal{R}$ then $(t,s) \in \mathcal{R}$.

(Transitivity) If $(s,t) \in \mathcal{R}$ and $(t,u) \in \mathcal{R}$ then $(s,u) \in \mathcal{R}$.

Example 2.4

Let $A = \{1,2,3,4\}$. The relation

$$\mathcal{R} = \{(1,1),(2,2),(3,3),(4,4),(1,4),(4,1),(2,4),(4,2),(1,2),(2,1)\}$$

is an equivalence relation on A. Check for yourself that reflexivity, symmetry, and transitivity all hold for $\mathcal{R}$.   ☐

The main result about an equivalence relation on A is that it induces a partition of A into disjoint sets:

THEOREM 2.3

Let $\mathcal{R}$ be an equivalence relation on a set A. Then A is a union of subsets A_α,

$$A = \bigcup_\alpha A_\alpha,$$

*with the following properties: If $a, b \in A$ then $(a, b) \in \mathcal{R}$ if and only if a and b are elements of the same A_α. The subsets A_α are nonempty and pairwise disjoint: $A_\alpha \cap A_{\alpha'} = \emptyset$ whenever $\alpha \neq \alpha'$. The sets A_α are called **equivalence classes**.*

PROOF   If $a \in A$ then define the subset $A(a)$ by

$$A(a) = \{b \in A : (a, b) \in \mathcal{R}\}.$$

Notice that, by the reflexive property of $\mathcal{R}, a \in A(a)$. So $A(a)$ is not empty. If $a, a' \in A$ and $A(a) \cap A(a') \neq \emptyset$ then there is at least one element common to the two sets: call it c. Then $c \in A(a)$ so that $(a, c) \in \mathcal{R}$. Also $c \in A(a')$ so that $(a', c) \in \mathcal{R}$. Now we invoke the symmetry property to conclude that $(c, a') \in \mathcal{R}$. Since $(a, c) \in \mathcal{R}$ and $(c, a') \in \mathcal{R}$, the transitivity property implies that $(a, a') \in \mathcal{R}$.

Now if b is any element of $A(a')$ then, by definition, $(a', b) \in \mathcal{R}$. We showed in the last paragraph that $(a, a') \in \mathcal{R}$. We conclude, by transitivity, that $(a, b) \in \mathcal{R}$. Hence $b \in A(a)$. Since b was an arbitrary element of $A(a')$, we have shown that $A(a') \subseteq A(a)$. The symmetry of the argument now gives that $A(a) \subseteq A(a')$. Thus $A(a) = A(a')$.

So we know that whenever two sets $A(a)$ and $A(a')$ intersect, they must be equal. Each of these sets is nonempty. And each $a \in A$ is in one of these sets (namely $A(a)$). This is what we wanted to prove.   ∎

Example 2.5

Refer to the first example. Notice that

$$A(1) = \{1, 2, 4\}, \quad A(2) = \{1, 2, 4\}, \quad A(3) = \{3\}, \quad A(4) = \{1, 2, 4\}.$$

Of course $A(1), A(2)$, and $A(4)$ are the same (as the Theorem predicts) because $(1, 2), (1, 4)$, and $(2, 4)$ are elements of $\mathcal{R}$. The equivalence relation $\mathcal{R}$ has partitioned A into the disjoint subsets $\{1, 2, 4\}$ and $\{3\}$. Also

$$A = \{1, 2, 4\} \cup \{3\}.$$ ☐

Example 2.6

Consider the set $\mathbb{N}$ of positive integers. Let $x, y \in \mathbb{N}$. We say that x is related to y if either $x \leq y$ and $y - x$ is divisible by 2 or $y < x$ and $x - y$ is divisible by 2. A moment's thought reveals that this means that two natural numbers are related if they are either both even or both odd.

Check for yourself that this is an equivalence relation (reflexivity is obvious; if x and y are both even/odd then so are y and x, giving symmetry; finally, write out the reasoning to verify transitivity).

The equivalence classes induced by this equivalence relation are $E = \{2, 4, 6, \ldots\}$ and $O = \{1, 3, 5, \ldots\}$. ☐

2.3   The Integers

Now we will apply the notion of an equivalence class to *construct* the integers (both positive and negative). There is an important point of knowledge to be noted here. For the sake of having a reasonable place to begin our work, we took the natural numbers $\mathbb{N} = \{1, 2, 3, \ldots\}$ as given. Since the natural numbers have been used for thousands of years to keep track of objects for barter, this is a plausible thing to do. Even people who know no mathematics accept the positive integers. However the number zero and the negative numbers are a different matter. It was not until the fifteenth century that the concepts of zero and negative numbers started to take hold — for they do not correspond to explicit collections of objects (five fingers or ten shoes) but rather to *concepts* (zero books is the lack of books; minus 4 pens means that we owe someone four pens). After some practice we get used to negative numbers, but explaining in words what they mean is always a bit clumsy.

It is much more satisfying, from the point of view of logic, to *construct* the integers from what we already have, that is, from the natural numbers. We proceed as follows. Let $A = \mathbb{N} \times \mathbb{N}$, the set of ordered pairs of natural numbers. We define a relation $\mathcal{R}$ on A and A as follows:

$$(a, b) \text{ is related to } (a', b') \text{ if } a + b' = a' + b.$$

THEOREM 2.4

The relation $\mathcal{R}$ is an equivalence relation.

PROOF   That (a, b) is related to (a, b) follows from the trivial identity $a + b = a + b$. Hence $\mathcal{R}$ is reflexive. Second, if (a, b) is related to (a', b') then $a + b' = a' + b$ hence $a' + b = a + b'$ (just reverse the equality) hence (a', b') is related to (a, b). So $\mathcal{R}$ is symmetric.

Finally, if (a, b) is related to (a', b') and (a', b') is related to (a'', b'') then we have

$$a + b' = a' + b \qquad \text{and} \qquad a' + b'' = a'' + b'.$$

Adding these equations gives

$$(a + b') + (a' + b'') = (a' + b) + (a'' + b').$$

Cancelling a' and b' from each side finally yields

$$a + b'' = a'' + b.$$

Thus (a, b) is related to (a'', b''). Therefore $\mathcal{R}$ is transitive. We conclude that $\mathcal{R}$ is an equivalence relation. ∎

Now our job is to understand the equivalence classes that are induced by $\mathcal{R}$. Let $(a, b) \in A$ and let $[(a, b)]$ be the corresponding equivalence class. If $b > a$ then we will denote this equivalence class by the integer $b - a$. For instance, the equivalence class $[(2, 7)]$ will be denoted by 5. Notice that if $(a', b') \in [(a, b)]$ then $a + b' = a' + b$ hence $b' - a' = b - a$. Therefore the integer that we choose to represent our equivalence class is *independent of the element of the equivalence class which is used to compute it.*

If $(a, b) \in A$ and $b = a$ then we let the symbol 0 denote the equivalence class $[(a, b)]$. Notice that if (a', b') is any other element of $[(a, b)]$ then it must be that $a + b' = a' + b$ hence $b' = a'$; therefore this definition is unambiguous.

If $(a, b) \in A$ and $a > b$ then we will denote the equivalence class $[(a, b)]$ by the symbol $-(a - b)$. For instance, we will denote the equivalence class $[(7, 5)]$ by the symbol -2. Once again, if (a', b') is related to (a, b) then the equation $a + b' = a' + b$ guarantees that our choice of symbol to represent $[(a, b)]$ is unambiguous.

Thus we have given our equivalence classes names, and these names *look just like* the names that we give to integers: there are positive integers, and negative ones, and zero. But we want to see that these objects *behave* like integers. (As you read on, use the mnemonic that the equivalence class $[(a, b)]$ stands for the integer $b - a$.)

First, do these new objects that we have constructed *add* correctly? Well, let $A = [(a, b)]$ and $A' = [(a', b')]$ be two equivalence classes. *Define* their sum to be $A + A' = [(a + a', b + b')]$. We must check that this is unambiguous. If $(\tilde{a}, \tilde{b})$ is related to (a, b) and $(\tilde{a}', \tilde{b}')$ is related to (a', b') then of course we know that

$$a + \tilde{b} = \tilde{a} + b$$

and

$$a' + \tilde{b}' = \tilde{a}' + b'.$$

Adding these two equations gives

$$(a + a') + (\tilde{b} + \tilde{b}') = (\tilde{a} + \tilde{a}') + (b + b')$$

hence $(a + a', b + b')$ is related to $(\tilde{a} + \tilde{a}', \tilde{b} + \tilde{b}')$. Thus adding two of our equivalence classes gives another equivalence class, as it should.

Example 2.7

To add 5 and 2 we first note that 5 is the equivalence class $[(2,7)]$ and 3 is the equivalence class $[(2,5)]$. We add them componentwise and find that the sum is $[(2+2, 7+5)] = [(4,12)]$. Which equivalence class is this answer? Looking back at our prescription for giving names to the equivalence classes, we see that this is the equivalence class that we called $12 - 4$ or 8. So we have rediscovered the fact that $5 + 3 = 8$.

Now let us add 4 and -9. The first of these is the equivalence class $[(3,7)]$ and the second is the equivalence class $[(13,4)]$. The sum is therefore $[(16,11)]$, and this is the equivalence class that we call $-(16 - 11)$ or -5. That is the answer that we would expect when we add 4 to -9.

Next, we add -12 and -5. Previous experience causes us to expect the answer to be -17. Now -12 is the equivalence class $[(19,7)]$ and -5 is the equivalence class $[(7,2)]$. The sum is $[(26,9)]$, which is the equivalence class that we call -17.

Finally, we can see in practice that our method of addition is unambiguous. Let us redo the second example using $[(6,10)]$ as the equivalence class represented by 4 and $[(15,6)]$ as the equivalence class represented by -9. Then the sum is $[(21,16)]$, and this is still the equivalence class -5, as it should be. ☐

REMARK 2.5 What is the point of this section? Everyone knows about negative numbers, so why go through this abstract construction? The reason is that until one sees this construction, negative numbers are just imaginary objects — placeholders if you will — which are a useful notation but which do not exist. Now they do exist. They are a collection of equivalence classes of pairs of natural numbers. This collection is equipped with certain arithmetic operations, such as addition, subtraction, and multiplication. We now discuss these last two. ∎

If $x = [(a, b)]$ and $x' = [(a', b')]$ are integers, we define their *difference* to be the equivalence class $[(a + b', b + a')]$; we denote this difference by $x - x'$. The unambiguity of this definition is treated in the exercises.

Example 2.8

We calculate $8 - 14$. Now $8 = [(1, 9)]$ and $14 = [(3, 17)]$. Therefore

$$8 - 14 = [(1 + 17, 9 + 3)] = [(18, 12)] = -6,$$

as expected.

As a second example, we compute $(-4) - (-8)$. Now

$$-4 - (-8) = [(6, 2)] - [(13, 5)] = [(6 + 5, 2 + 13)] = [(11, 15)] = 4.$$

REMARK 2.6 When we first learn that $(-4) - (-8) = (-4) + 8 = 4$, the explanation is a bit mysterious: why is "minus a minus equal to a plus"? Now there is no longer any mystery: this property follows *from our construction* of the number system $\mathbb{Z}$.

Finally, we turn to multiplication. If $x = [(a, b)]$ and $x' = [(a', b')]$ are integers then we define their product by the formula

$$x \cdot x' = [(a \cdot b' + b \cdot a', a \cdot a' + b \cdot b')].$$

This definition may be a surprise. Why did we not define $x \cdot x'$ to be $[(a \cdot a', b \cdot b')]$? There are several reasons: first of all, the latter definition would give the wrong answer; moreover, it is not unambiguous (different representatives of x and x' would give different answers). If you recall that we think of $[(a, b)]$ as representing $b - a$ and $[(a', b')]$ as representing $b' - a'$ then the product should be the equivalence class that represents $(b - a) \cdot (b' - a')$. That is the motivation behind our definition.

The unambiguity of the given definition of multiplication of integers is treated in the exercises. We proceed now to an example.

Example 2.9

We compute the product of -3 and -6. Now

$$(-3) \cdot (-6) = [(5, 2)] \cdot [(9, 3)] = [(5 \cdot 3 + 2 \cdot 9, 5 \cdot 9 + 2 \cdot 3)] = [(33, 51)] = 18,$$

which is the expected answer.

As a second example, we multiply -5 and 12. We have

$$-5 \cdot 12 = [(7, 2)] \cdot [(1, 13)] = [(7 \cdot 13 + 2 \cdot 1, 7 \cdot 1 + 2 \cdot 13)] = [(93, 33)] = -60.$$

Finally, we show that 0 times any integer A equals zero. Let $A = [(a, b)]$. Then

$$0 \cdot A = [(1, 1)] \cdot [(a, b)] = [(1 \cdot b + 1 \cdot a, 1 \cdot a + 1 \cdot b)]$$
$$= [(a + b, a + b)] = 0.$$

REMARK 2.7   Notice that one of the pleasant by-products of our construction of the integers is that we no longer have to give artificial explanations for why the product of two negative numbers is a positive number or why the product of a negative number and a positive number is negative. These properties instead follow automatically from our construction.   ∎

Of course we will not discuss division for integers; in general, division of one integer by another makes no sense *in the universe of the integers*. More will be said about this in the exercises.

In the rest of this book we will follow the standard mathematical custom of denoting the set of all integers by the symbol $\mathbb{Z}$. We will write the integers not as equivalence classes, but in the usual way as $\ldots -3, -2, -1, 0, 1, 2, 3, \ldots$. The equivalence classes are a device that we used to *construct* the integers. Now that we have them, we may as well write them in the simple, familiar fashion.

In an exhaustive treatment of the construction of $\mathbb{Z}$, we would prove that addition and multiplication are commutative and associative, prove the distributive law, and so forth. But the purpose of this section is to demonstrate modes of logical thought rather than to be exhaustive. We shall say more about some of the elementary properties of the integers in the exercises.

2.4   The Rational Numbers

In this section we use the integers, together with a construction using equivalence classes, to build the rational numbers. Let A be the set $\mathbb{Z} \times (\mathbb{Z} \setminus \{0\})$. Here the symbol $\setminus$ stands for "subtraction of sets": $\mathbb{Z} \setminus \{0\}$ denotes the set of all elements of $\mathbb{Z}$ *except* 0 (see Section 1.6). In other words, A is the set of ordered pairs (a, b) of integers subject to the condition that $b \neq 0$. [*Think of this ordered pair as "representing" the fraction a/b.*] We definitely want it to be the case that certain ordered pairs represent the same number. For instance,

$$\tfrac{1}{2} \text{ should be the same number as } \tfrac{3}{6}.$$

This motivates our equivalence relation. Declare (a, b) to be related to (a', b') if $a \cdot b' = a' \cdot b$. [*Here we are thinking that the fraction a/b should equal the fraction a'/b' precisely when $a \cdot b' = a' \cdot b$.*]

Is this an equivalence relation? Obviously the pair (a, b) is related to itself, since $a \cdot b = a \cdot b$. Also the relation is symmetric: if (a, b) and (a', b') are pairs and $a \cdot b' = a' \cdot b$, then $a' \cdot b = a \cdot b'$. Finally, if (a, b) is related to (a', b') and (a', b') is related to (a'', b'') then we have both

$$a \cdot b' = a' \cdot b \qquad \text{and} \qquad a' \cdot b'' = a'' \cdot b'.$$

Multiplying the left sides of these two equations together and the right sides together gives

$$(a \cdot b') \cdot (a' \cdot b'') = (a' \cdot b) \cdot (a'' \cdot b').$$

If $a' = 0$ then it follows immediately that both a and a'' must be zero. The three pairs $(a, b), (a', b')$, and (a'', b'') are then equivalent and there is nothing to prove. So we may assume that $a' \neq 0$. We know *a priori* that $b' \neq 0$; therefore we may cancel common terms in the last equation to obtain

$$a \cdot b'' = b \cdot a''.$$

Thus (a, b) is related to (a'', b''), and our relation is transitive.

The resulting collection of equivalence classes will be called the set of *rational numbers*, and we shall denote this set with the symbol $\mathbb{Q}$.

Example 2.10
The equivalence class $[(4, 12)]$ contains all of the pairs $(4, 12), (1, 3), (-2, -6)$. (Of course it contains infinitely many other pairs as well.) This equivalence class represents the fraction $4/12$ which we sometimes also write as $1/3$ or $-2/(-6)$.

□

If $[(a, b)]$ and $[(a', b')]$ are rational numbers then we define their *product* to be the rational number

$$[(a \cdot a', b \cdot b')].$$

This is well defined, for if (a, b) is related to $(\tilde{a}, \tilde{b})$ and (a', b') is related to $(\tilde{a}', \tilde{b}')$ then we have the equations

$$a \cdot \tilde{b} = \tilde{a} \cdot b \quad \text{and} \quad a' \cdot \tilde{b}' = \tilde{a}' \cdot b'.$$

Multiplying together the left and right sides we obtain

$$(a \cdot \tilde{b}) \cdot (a' \cdot \tilde{b}') = (\tilde{a} \cdot b) \cdot (\tilde{a}' \cdot b').$$

Rearranging, we have

$$(a \cdot a') \cdot (\tilde{b} \cdot \tilde{b}') = (\tilde{a} \cdot \tilde{a}') \cdot (b \cdot b').$$

But this says that the product of $[(a, b)]$ and $[(a', b')]$ is related to the product of $[(\tilde{a}, \tilde{b})]$ and $[(\tilde{a}', \tilde{b}')]$. So multiplication is unambiguous.

Example 2.11
The product of the two rational numbers $[(3, 8)]$ and $[(-2, 5)]$ is

$$[(3 \cdot (-2), 8 \cdot 5)] = [(-6, 40)] = [(-3, 20)].$$

This is what we expect: the product of $3/8$ and $-2/5$ is $-3/20$. ▯

If $q = [(a, b)]$ and $r = [(c, d)]$ are rational numbers and if r is not zero (that is, $[(c, d)]$ is not the equivalence class zero — in other words, $c \neq 0$) then we define the quotient q/r to be the equivalence class

$$[(ad, bc)].$$

We leave it to you to check that this operation is well defined.

Example 2.12
The quotient of the rational number $[(4, 7)]$ by the rational number $[(3, -2)]$ is, by definition, the rational number

$$[(4 \cdot (-2), 7 \cdot 3)] = [(-8, 21)].$$

This is what we expect: the quotient of $4/7$ by $-3/2$ is $-8/(21)$. ▯

How should we add two rational numbers? We could try declaring $[(a, b)] + [(a', b')]$ to be $[(a + a', b + b')]$, but this will not work (think about the way that we usually add fractions). Instead we define

$$[(a, b)] + [(a', b')] = [(a \cdot b' + a' \cdot b, b \cdot b')].$$

That this definition is unambiguous is left for the exercises. We turn instead to an example.

Example 2.13
The sum of the rational numbers $[(3, -14)]$ and $[(9, 4)]$ is given by

$$[(3 \cdot 4 + 9 \cdot (-14), (-14) \cdot 4)] = [(-114, -56)] = [(57, 28)].$$

This coincides with the usual way that we add fractions:

$$-\frac{3}{14} + \frac{9}{4} = \frac{57}{28}.$$ ▯

Notice that the equivalence class $[(0, 1)]$ is the rational number that we usually denote by 0. It is the additive identity, for if $[(a, b)]$ is another rational number then

$$[(0, 1)] + [(a, b)] = [(0 \cdot b + a \cdot 1, 1 \cdot b)] = [(a, b)].$$

A similar argument shows that $[(0, 1)]$ times any rational number gives $[(0, 1)]$ or 0.

Of course the concept of subtraction is really just a special case of addition (that is, $a - b$ is the same thing as $a + (-b)$). So we shall say nothing further about subtraction.

In practice we will write rational numbers in the traditional fashion:

$$\frac{2}{5}, \quad \frac{-19}{3}, \quad \frac{22}{2}, \quad \frac{24}{4}, \quad \ldots .$$

In mathematics it is generally not wise to write rational numbers in mixed form, such as $2\frac{3}{5}$, because the juxtaposition of two numbers could easily be mistaken for multiplication. Instead we would write this quantity as the improper fraction $13/5$.

DEFINITION 2.14 *A set S is called a **field** if it is equipped with a binary operation (usually called addition and denoted "$+$") and a second binary operation (called multiplication and denoted "$\cdot$") such that the following axioms are satisfied:*

A1. *S is closed under addition: if $x, y \in S$ then $x + y \in S$.*

A2. *Addition is commutative: if $x, y \in S$ then $x + y = y + x$.*

A3. *Addition is associative: if $x, y, z \in S$ then $x + (y + z) = (x + y) + z$.*

A4. *There exists an element, called 0, in S that is an additive identity: if $x \in S$ then $0 + x = x$.*

A5. *Each element of S has an additive inverse: if $x \in S$ then there is an element $-x \in S$ such that $x + (-x) = 0$.*

M1. *S is closed under multiplication: if $x, y \in S$ then $x \cdot y \in S$.*

M2. *Multiplication is commutative: if $x, y \in S$ then $x \cdot y = y \cdot x$.*

M3. *Multiplication is associative: if $x, y, z \in S$ then $x \cdot (y \cdot z) = (x \cdot y) \cdot z$.*

M4. *There exists an element, called 1, that is a multiplicative identity: if $x \in S$ then $x \cdot 1 = x$.*

M5. *Each nonzero element of S has a multiplicative inverse: if $0 \neq x \in S$ then there is an element $x^{-1} \in S$ such that $x \cdot (x^{-1}) = 1$. The element x^{-1} is sometimes denoted $1/x$.*

D1. *Multiplication distributes over addition: if $x, y, z \in S$ then $x \cdot (y + z) = x \cdot y + x \cdot z$.*

Eleven axioms is a lot to digest all at once, but in fact these are all familiar properties of addition and multiplication of rational numbers that we use every day: the set $\mathbb{Q}$, with the usual notions of addition and multiplication, forms a field. The integers, by contrast, do not: nonzero elements of $\mathbb{Z}$ (except 1 and -1) do not have multiplicative inverses *in the integers*.

Let us now consider some consequence of the field axioms.

THEOREM 2.8

Any field has the following properties:

1. *If $z + x = z + y$ then $x = y$.*
2. *If $x + z = 0$ then $z = -x$ (the additive inverse is unique).*
3. *$-(-y) = y$.*
4. *If $y \neq 0$ and $y \cdot x = y \cdot z$ then $x = z$.*
5. *If $y \neq 0$ and $y \cdot z = 1$ then $z = y^{-1}$ (the multiplicative inverse is unique).*
6. *$\left(x^{-1}\right)^{-1} = x$.*
7. *$0 \cdot x = 0$.*
8. *If $x \cdot y = 0$ then either $x = 0$ or $y = 0$.*
9. *$(-x) \cdot y = -(x \cdot y) = x \cdot (-y)$.*
10. *$(-x) \cdot (-y) = x \cdot y$.*

PROOF   These are all familiar properties of the rationals, but now we are considering them for an arbitrary field. We prove just a few to illustrate the logic. The proofs of the others are assigned as exercises.

To prove (1) we write

$$z + x = z + y \Rightarrow (-z) + (z + x) = (-z) + (z + y)$$

and now Axiom A3 yields that this implies

$$((-z) + z) + x = ((-z) + z) + y.$$

This can be rewritten (using Axioms A2, A5) as

$$0 + x = 0 + y$$

and hence, by Axiom A4,

$$x = y.$$

To prove (7), we observe that

$$0 \cdot x = (0 + 0) \cdot x,$$

which by Axiom M2 equals

$$x \cdot (0 + 0).$$

By Axiom D1 the last expression equals

$$x \cdot 0 + x \cdot 0,$$

which by Axiom M2 equals $0 \cdot x + 0 \cdot x$. Thus we have derived the equation

$$0 \cdot x = 0 \cdot x + 0 \cdot x.$$

Axioms A4 and A2 let us rewrite the left side so that we have

$$0 \cdot x + 0 = 0 \cdot x + 0 \cdot x.$$

Finally, part (1) of the present theorem (which we have already proved) yields that

$$0 = 0 \cdot x,$$

which is the desired result.

To prove (8), we suppose that $x \neq 0$. In this case x has a multiplicative inverse x^{-1} and we multiply both sides of our equation by this element:

$$x^{-1} \cdot (x \cdot y) = x^{-1} \cdot 0.$$

By Axiom M3, the left side can be rewritten and we have

$$(x \cdot x^{-1}) \cdot y = x^{-1} \cdot 0.$$

Next, we rewrite the right side using Axiom M2:

$$(x \cdot x^{-1}) \cdot y = 0 \cdot x^{-1}.$$

Now Axiom M5 allows us to simplify the left side:

$$1 \cdot y = 0 \cdot x^{-1}.$$

We further simplify the left side using Axioms M2, M4 and the right side using Part (7) of the present theorem (which we just proved) to obtain:

$$y = 0.$$

Thus we see that if $x \neq 0$ then $y = 0$. But this is logically equivalent with $x = 0$ or $y = 0$, as we wished to prove. [If you have forgotten why these statements are logically equivalent, write a truth table.]   ∎

DEFINITION 2.15   *Let A be a set. We shall say that A is **ordered** if there is a relation $\mathcal{R}$ on A and A satisfying the following properties:*

1. *If $a \in A$ and $b \in A$ then one and only one of the following holds:*
    $(a, b) \in \mathcal{R}$ or $(b, a) \in \mathcal{R}$ or $a = b$.   not   symm; trans

2. *If a, b, c are elements of A and $(a, b) \in \mathcal{R}$ and $(b, c) \in \mathcal{R}$ then $(a, c) \in \mathcal{R}$.*

*We shall call the relation $\mathcal{R}$ an **order** on A.*

Rather than write an ordering relation as $(a, b) \in \mathcal{R}$, it is usually more convenient to write it as $a < b$. The notation $b > a$ means the same thing as $a < b$.

Example 2.16

The integers $\mathbb{Z}$ form an ordered set with the usual ordering. We can make this ordering precise by saying that $x < y$ if $y - x$ is a positive integer. For instance,

$$6 < 8 \quad \text{because} \quad 8 - 6 = 2 > 0.$$

Likewise,

$$-5 < -1 \quad \text{because} \quad -1 - (-5) = 4 > 0.$$

Observe that the same ordering works on the rational numbers. ☐

If A is an ordered set and a, b are elements then we often write $a \leq b$ to mean that *either $a = b$ or $a < b$.*

When a field has an ordering that is compatible with the field operations then a richer structure results:

DEFINITION 2.17 *A field F is called an **ordered field** if F has an ordering $<$ that satisfies the following addition properties:*

1. *If $x, y, z \in F$ and $y < z$ then $x + y < x + z$.*
2. *If $x, y \in F, x > 0$, and $y > 0$ then $x \cdot y > 0$.*

Again, these are familiar properties of the rational numbers: $\mathbb{Q}$ forms an ordered field.

THEOREM 2.9

*Any **ordered** field has the following properties:*

1. *If $x > 0$ and $z < y$ then $x \cdot z < x \cdot y$.*
2. *If $x < 0$ and $z < y$ then $x \cdot z > x \cdot y$.*
3. *If $x > 0$ then $-x < 0$. If $x < 0$ then $-x > 0$.*
4. *If $0 < y < x$ then $0 < 1/x < 1/y$.*
5. *If $x \neq 0$ then $x^2 > 0$.*
6. *If $0 < x < y$ then $x^2 < y^2$.*

PROOF   Again we prove just a few of these statements and leave the rest as exercises.

To prove (1), observe that the property (1) of ordered fields together with our hypothesis implies that

$$(-z) + z < (-z) + y.$$

Thus, using (A2), we see that $y - z > 0$. Since $x > 0$, property (2) of ordered fields gives

$$x \cdot (y - z) > 0.$$

Finally,

$$x \cdot y = x \cdot [(y - z) + z] = x \cdot (y - z) + x \cdot z > 0 + x \cdot z$$

(by property (1) again). In conclusion,

$$x \cdot y > x \cdot z.$$

To prove (3), begin with the equation

$$0 = -x + x.$$

Since $x > 0$, the right side is greater than $-x$. Thus $0 > -x$ as claimed. The proof of the other statement of (3) is similar.

To prove (5), we consider two cases. If $x > 0$ then $x^2 \equiv x \cdot x$ is positive by property (2) of ordered fields. If $x < 0$ then $-x > 0$ (by part (3) of the present theorem, which we just proved) hence $(-x) \cdot (-x) > 0$. But part (10) of the last theorem guarantees that $(-x) \cdot (-x) = x \cdot x$ hence $x \cdot x > 0$.   ∎

We conclude this section by recording an inadequacy of the field of rational numbers; this will serve in part as motivation for learning about the real numbers in the next section.

THEOREM 2.10
There is no positive rational number q such that $q^2 = q \cdot q = 2$.

PROOF   Seeking a contradiction, suppose that there is such a q. Write q in lowest terms as

$$q = \frac{a}{b},$$

with a and b integers greater than zero. This means that the numbers a and b have no common divisors except 1. The equation $q^2 = 2$ can then be written as

$$a^2 = 2 \cdot b^2.$$

Since 2 divides the right side of this last equation, it follows that 2 divides the left side. But 2 can divide a^2 only if 2 divides a (because 2 is prime). We write $a = 2 \cdot \alpha$ for some positive integer α. But then the last equation becomes

$$4 \cdot \alpha^2 = 2 \cdot b^2.$$

Simplifying yields that

$$2 \cdot a^2 = b^2.$$

Since 2 divides the left side, we conclude that 2 must divide the right side. But 2 can divide b^2 only if 2 divides b.

This is our contradiction: we have argued that 2 divides a *and* that 2 divides b. But a and b were assumed to *have no common divisors*. We conclude that the rational number q cannot exist. ∎

In fact it turns out that a positive integer can be the square of a rational number if and only if it is the square of a positive integer. This assertion is explored in Exercise 36. It is a special case of a more general phenomenon in number theory known as Gauss's lemma.

2.5 The Real Numbers

Now that we are accustomed to the notion of equivalence classes, the construction of the integers and of the rational numbers seems fairly natural. In fact, equivalence classes provide a precise language for declaring certain objects to be equal. We can now use the integers and the rationals as we always have done, with the added confidence that they are not simply a useful notation but that they have been *constructed*.

We turn next to the real numbers. We know from calculus that for many purposes the rational numbers are inadequate. It is important to work in a number system that is closed with respect to the operations we shall perform. While the rationals are closed under the usual arithmetic operations, they are not closed under the operation of taking *limits*. For instance, the sequence of rational numbers $3, 3.1, 3.14, 3.141, \ldots$ consists of terms that seem to be getting closer and closer together, *seem* to tend to some limit, and yet there is no rational number that will serve as a limit (of course it turns out that the limit is π — an "irrational" number).

We will now deal with the real number system, a system that contains all limits of sequences of rational numbers (as well as all limits of sequences of real numbers!). In fact our plan will be as follows: in this section we shall discuss all the requisite properties of the reals. The actual construction of the reals is rather complicated, and we shall put that in an Appendix to this chapter.

DEFINITION 2.18 *Let A be an ordered set and X a subset of A. The set X is called **bounded above** if there is an element $b \in A$ such that $x \le b$ for all $x \in X$. We call the element b an **upper bound** for the set X.*

Example 2.19

Let $A = \mathbb{Q}$ with the usual ordering. The set $X = \{x \in \mathbb{Q} : 2 < x < 4\}$ is bounded above. For example, 15 is an upper bound for X. So are the numbers 12 and 4. It is interesting to observe that no element of this particular X can be an upper bound for X. The number 4 is a good candidate, but 4 is not an element of X. In fact, if $b \in X$ then $(b+4)/2 \in X$ and $b < (b+4)/2$, so b could not be an upper bound for X. ꗃ

It turns out that the most convenient way to formulate the notion that the real numbers have "no holes" (i.e., that all sequences that seem to be converging actually have something to converge to) is in terms of upper bounds.

DEFINITION 2.20   *Let A be an ordered set and X a subset of A. An element $b \in A$ is called a **least upper bound** (or **supremum**) for X if b is an upper bound for X and there is no upper bound b' for X that is less than b. We write $b = \mathrm{lub}\, X$ or $b = \sup X$.*
     By its very definition, if a least upper bound exists then it is unique.

Example 2.21

In the last example, we considered the set X of rational numbers strictly between 2 and 4. We observed there that 4 is the least upper bound for X. Note that this least upper bound is not an element of the set X.
     The set $Y = \{y \in \mathbb{Z} : -9 \leq y \leq 7\}$ has least upper bound 7. In this case, the least upper bound *is* an element of the set Y. ꗃ

Notice that we may define a lower bound for a subset of an ordered set in a fashion similar to that for an upper bound: $l \in A$ is a lower bound for $X \subseteq A$ if $x \geq l$ for all $x \in X$. A *greatest lower bound* (or *infimum*) for X is then defined to be a lower bound l such that there is no lower bound l' with $l' > l$. We write $l = \mathrm{glb}\, X$ or $l = \inf X$.

Example 2.22

The set X in the last two examples has lower bounds -20, 0, 1, 2, for instance. The greatest lower bound is 2, which is *not* an element of the set.
     The set Y in the last example has lower bounds -53, -22, -10, -9, to name just a few. The number -9 is the greatest lower bound. It *is* an element of Y. ꗃ

The purpose that the real numbers will serve for us is as follows: they will contain the rationals, they will still be an ordered field, and ***every subset that***

has an upper bound will have a least upper bound. We formulate this as a theorem.

THEOREM 2.11
There exists an ordered field $\mathbb{R}$ that (i) contains $\mathbb{Q}$ and (ii) has the property that any nonempty subset of $\mathbb{R}$ which has an upper bound has a least upper bound.

The last property described in this theorem is called the Least Upper Bound Property of the real numbers. As mentioned previously, this theorem will be proved in the Appendix to the chapter. Now we begin to realize why it is so important to *construct* the number systems that we will use. We are endowing $\mathbb{R}$ with a great many properties. Why do we have any right to suppose that there exists a set with all these properties? We must produce one!

Let us begin to explore the richness of the real numbers. The next theorem states a property that is certainly not shared by the rationals (see Theorem 2.10). It is fundamental in its importance.

THEOREM 2.12
Let x be a real number such that $x > 0$. Then there is a positive real number y such that $y^2 = y \cdot y = x$.

PROOF We will use throughout this proof the fact (see Part (6) of Theorem 2.9) that if $0 < a < b$ then $a^2 < b^2$.
Let

$$S = \{s \in \mathbb{R} : s > 0 \text{ and } s^2 < x\}.$$

Then S is not empty since $x/2 \in S$ if $x < 2$ and $1 \in S$ otherwise. Also, S is bounded above since $x + 1$ is an upper bound for S. By Theorem 2.11, the set S has a least upper bound. Call it y. Obviously $0 < \min\{x/2, 1\} \leq y$; hence y is positive. We claim that $y^2 = x$. To see this, we eliminate the other two possibilities.
If $y^2 < x$ then set $\epsilon = (x - y^2)/[4(x + 1)]$. Then $\epsilon > 0$ and

$$
\begin{aligned}
(y + \epsilon)^2 &= y^2 + 2 \cdot y \cdot \epsilon + \epsilon^2 \\
&= y^2 + 2 \cdot y \cdot \frac{x - y^2}{4(x + 1)} + \frac{x - y^2}{4(x + 1)} \cdot \frac{x - y^2}{4(x + 1)} \\
&< y^2 + 2 \cdot y \cdot \frac{x - y^2}{4y} + \frac{x - y^2}{4(x + 1)} \cdot \frac{x - y^2}{4(x + 1)} \\
&< y^2 + \frac{x - y^2}{2} + \frac{x - y^2}{4} \cdot \frac{x}{4x} \\
&< y^2 + (x - y^2) \\
&= x.
\end{aligned}
$$

Thus $y + \epsilon \in S$, and y cannot be an upper bound for S. This contradiction tells us that $y^2 \not< x$.

Similarly, if it were the case that $y^2 > x$ then we set $\epsilon = (y^2 - x)/[4(x+1)]$. A calculation like the one we just did (see Exercise 27) then shows that if $s \geq y - \epsilon$ then $s^2 \geq x$. Hence $y - \epsilon$ is also an upper bound for S, and y is therefore not the *least* upper bound. This contradiction shows that $y^2 \not> x$.

The only remaining possibility is that $y^2 = x$. ∎

A similar proof shows that if n is a positive integer and x a positive real number then there is a positive real number y such that $y^n = x$. Exercise 35 asks you to provide the details.

We next use the Least Upper Bound Property of the Real Numbers to establish two important qualitative properties of the Real Numbers.

THEOREM 2.13

The set $\mathbb{R}$ of real numbers satisfies the Archimedean Property:

> *Let a and b be positive real numbers. Then there is a natural number n such that $na > b$.*

The set $\mathbb{Q}$ of rational numbers satisfies the following Density Property:

> *Let $c < d$ be real numbers. Then there is a rational number q with $c < q < d$.*

PROOF Suppose the Archimedean Property to be false. Then $S = \{na : n \in \mathbb{N}\}$ has b as an upper bound. Therefore S has a finite supremum β. Since $a > 0$, $\beta - a < \beta$. So $\beta - a$ is not an upper bound for S, and there must be a natural number n' such that $n' \cdot a > \beta - a$. But then $(n' + 1)a > \beta$, and β cannot be the supremum for S. This contradiction proves the first assertion.

For the second property, let $\lambda = d - c > 0$. By the Archimedean Property, choose a positive integer N such that $N \cdot \lambda > 1$. Again the Archimedean Property gives a natural number P such that $P > N \cdot c$ and another Q such that $Q > -N \cdot c$. Thus we see that Nc falls between the integers $-Q$ and P; therefore there must be an integer M between $-Q$ and P such that

$$M - 1 \leq Nc < M.$$

Thus $c < M/N$. Also

$$M \leq Nc + 1 \quad \text{hence} \quad \frac{M}{N} \leq c + \frac{1}{N} < c + \lambda = d.$$

So M/N is a rational number lying between c and d. ∎

Recall that in Example 1.51 in Section 1.8 we established that the set of all decimal representations of numbers is uncountable. It follows that the set of all real numbers is uncountable. In fact the same proof shows that the set of all real numbers in the interval $(0, 1)$, or in any nonempty open interval (c, d), is uncountable.

The set $\mathbb{R}$ of real numbers is uncountable, yet the set $\mathbb{Q}$ of rational numbers is countable. It follows that the set $\mathbb{R} \setminus \mathbb{Q}$ of *irrational* numbers is uncountable. In particular, it is nonempty. Thus we may see with very little effort that there exist a great many real numbers that cannot be expressed as a quotient of integers. However, given any particular real number (such as π or e or $\sqrt[5]{2}$), it can be quite difficult to see whether or not it is irrational.

We conclude by recalling the "absolute value" notation:

DEFINITION 2.23 *Let x be a real number. We define*

$$|x| = \begin{cases} x & if \quad x > 0 \\ 0 & if \quad x = 0 \\ -x & if \quad x < 0 \end{cases}$$

It is left as an exercise for you to verify the important *triangle inequality*:

$$|x + y| \leq |x| + |y|.$$

2.6 The Complex Numbers

When we first learn about the complex numbers, the most troublesome point is the very beginning: "Let's pretend that the number -1 has a square root. Call it i." What gives us the right to "pretend" in this fashion? The answer is that we have no such right. If -1 has a square root, we should be able to construct a number system in which that is the case. That is what we shall do in this section.

DEFINITION 2.24 *The system of* **complex numbers**, *denoted by the symbol* $\mathbb{C}$, *consists of all ordered pairs (a, b) of real numbers. We add two complex numbers (a, b) and (a', b') by the formula*

$$(a, b) + (a', b') = (a + a', b + b').$$

We multiply two complex numbers by the formula

$$(a, b) \cdot (a', b') = (a \cdot a' - b \cdot b', a \cdot b' + a' \cdot b).$$

REMARK 2.14 If you are puzzled by this definition of multiplication, do not worry. In a few moments you will see that it gives rise to the notion of multiplication of complex numbers that you are used to. ∎

Example 2.25

Let $z = (3, -2)$ and $w = (4, 7)$ be two complex numbers. Then

$$z + w = (3, -2) + (4, 7) = (3 + 4, -2 + 7) = (7, 5).$$

Also

$$z \cdot w = (3, -2) \cdot (4, 7) = (3 \cdot 4 - (-2) \cdot 7, 3 \cdot 7 + 4 \cdot (-2)) = (26, 13). \quad \Box$$

As usual, we ought to check that addition and multiplication are commutative and associative, that multiplication distributes over addition, and so forth. We shall leave these tasks to the exercises. Instead we develop some of the crucial properties of our new number system.

THEOREM 2.15

The following properties hold for the number system $\mathbb{C}$.

1. *The number* $1 \equiv (1, 0)$ *is the multiplicative identity:* $1 \cdot z = z$ *for any* $z \in \mathbb{C}$.

2. *The number* $0 \equiv (0, 0)$ *is the additive identity:* $0 + z = z$ *for any* $z \in \mathbb{C}$.

3. *Each complex number* $z = (x, y)$ *has an additive inverse* $-z = (-x, -y)$: *it holds that* $z + -z = 0$.

4. *The number* $i \equiv (0, 1)$ *satisfies* $i \cdot i = -1$; *in other words,* i *is a square root of* -1.

PROOF   These are direct calculations, but it is important for us to work out these facts.

First, let $z = (x, y)$ be any complex number. Then

$$1 \cdot z = (1, 0) \cdot (x, y) = (1 \cdot x - 0 \cdot y, 1 \cdot y + x \cdot 0) = (x, y) = z.$$

This proves the first assertion.

For the second, we have

$$0 + z = (0, 0) + (x, y) = (0 + x, 0 + y) = (x, y) = z.$$

With z as above, set $-z = (-x, -y)$. Then

$$z + (-z) = (x, y) + (-x, -y) = (x + (-x), y + (-y)) = (0, 0) = 0.$$

Finally, we calculate

$$i \cdot i = (0, 1) \cdot (0, 1) = (0 \cdot 0 - 1 \cdot 1, 0 \cdot 1 + 0 \cdot 1) = (-1, 0) = -1.$$

Thus, as asserted, i is a square root of -1.   ∎

PROPOSITION 2.16

If $z \in \mathbb{C}, z \neq 0$, then there is a complex number w such that $z \cdot w = 1$.

PROOF Write $z = (x, y)$ and set

$$w = \left(\frac{x}{x^2 + y^2}, \frac{-y}{x^2 + y^2} \right).$$

Since $z \neq 0$, this definition makes sense. Then it is straightforward to verify that $z \cdot w = 1$. ∎

Thus every nonzero complex number has a multiplicative inverse. The other field axioms for $\mathbb{C}$ are easy to check. We conclude that the number system $\mathbb{C}$ forms a field. You will prove in the exercises that it is not possible to order this field. If α is a real number then we associate α with the complex number $(\alpha, 0)$. In this way, we can think of the real numbers as a *subset* of the complex numbers. In fact, the real field $\mathbb{R}$ is a *subfield* of the complex field $\mathbb{C}$. This means that if $\alpha, \beta \in \mathbb{R}$ and $(\alpha, 0), (\beta, 0)$ are the corresponding elements in $\mathbb{C}$ then $\alpha + \beta$ corresponds to $(\alpha + \beta, 0)$ and $\alpha \cdot \beta$ corresponds to $(\alpha, 0) \cdot (\beta, 0)$. These assertions are explored more thoroughly in the exercises.

With the remarks in the preceding paragraph we can sometimes ignore the distinction between the real numbers and the complex numbers. For example, we can write

$$5 \cdot i$$

and understand that it means $(5, 0) \cdot (0, 1) = (0, 5)$. Likewise, the expression

$$5 \cdot 1$$

can be interpreted as $5 \cdot 1 = 5$ or as $(5, 0) \cdot (1, 0) = (5, 0)$ without any danger of ambiguity.

THEOREM 2.17

Every complex number can be written in the form $a + b \cdot i$, where a and b are real numbers. In fact, if $z = (x, y) \in \mathbb{C}$ then

$$z = x + y \cdot i.$$

PROOF With the identification of real numbers as a subfield of the complex numbers, we have that

$$x + y \cdot i = (x, 0) + (y, 0) \cdot (0, 1) = (x, 0) + (0, y) = (x, y) = z$$

as claimed. ∎

Now that we have constructed the complex number field, we will adhere to the usual custom of writing complex numbers as $z = a + b \cdot i$ or, more simply, $a + bi$. We call a the *real part* of z, denoted by Re z, and b the *imaginary part* of z, denoted by Im z. We have

$$(a + bi) + (a' + b'i) = (a + a') + (b + b')i$$

and

$$(a + bi) \cdot (a' + b'i) = (a \cdot a' - b \cdot b') + (a \cdot b' + a' \cdot b)i.$$

If $z = a + bi$ is a complex number then we define its *complex conjugate* to be the number $\bar{z} = a - bi$. We record some elementary facts about the complex conjugate:

PROPOSITION 2.18

If z, w are complex numbers then

1. $\overline{z + w} = \bar{z} + \bar{w}$;
2. $\overline{z \cdot w} = \bar{z} \cdot \bar{w}$;
3. $z + \bar{z} = 2 \cdot Re\, z$;
4. $z - \bar{z} = 2 \cdot i \cdot Im\, z$;
5. $z \cdot \bar{z} \geq 0$, *with equality holding if and only if $z = 0$.*

PROOF   Write $z = a + bi, w = c + di$. Then

$$\overline{z + w} = \overline{(a + c) + (b + d)i}$$
$$= (a + c) - (b + d)i$$
$$= (a - bi) + (c - di)$$
$$= \bar{z} + \bar{w}.$$

This proves (1). Assertions (2), (3), (4) are proved similarly. For (5), notice that

$$z \cdot \bar{z} = (a + bi) \cdot (a - bi) = a^2 + b^2 \geq 0.$$

Clearly equality holds if and only if $a = b = 0$.   ∎

The expression $|z|$ is defined to be the nonnegative square root of $z \cdot \bar{z}$. It is called the *modulus* of z and plays the same role for the complex field that absolute value plays for the real field. The modulus has the following properties:

PROPOSITION 2.19

If $z, w \in \mathbb{C}$ then

1. $|z| = |\bar{z}|$;
2. $|z \cdot w| = |z| \cdot |w|$;
3. $|\mathrm{Re}\ z| \leq |z|, |\mathrm{Im}\ z| \leq |z|$;
4. $|z + w| \leq |z| + |w|$.

PROOF Write $z = a + bi, w = c + di$. Then (1), (2), (3) are immediate. For (4) we calculate that

$$
\begin{aligned}
|z + w|^2 &= (z + w) \cdot (\overline{z + w}) \\
&= z \cdot \bar{z} + z \cdot \bar{w} + w \cdot \bar{z} + w \cdot \bar{w} \\
&= |z|^2 + 2\mathrm{Re}\ (z \cdot \bar{w}) + |w|^2 \\
&\leq |z|^2 + 2|z \cdot \bar{w}| + |w|^2 \\
&= |z|^2 + 2|z| \cdot |w| + |w|^2 \\
&= (|z| + |w|)^2.
\end{aligned}
$$

Taking square roots proves (4). ∎

Observe that if z is real then $z = a + 0i$ and the modulus of z equals the absolute value of a. Likewise, if $z = 0 + bi$ is pure imaginary then the modulus of z equals the absolute value of b. In particular, the third part of the Proposition reduces, in the real case, to the triangle inequality

$$
|x + y| \leq |x| + |y|.
$$

Exercises

2.1 Consider the following alternate form of the Principle of Induction: Let Q be a property that may or may not hold for all of the natural numbers $\mathbb{N}$. Assume that 1 has the property Q, and that whenever j has the property Q for $1 \leq j < n$ then n has the property Q; then it follows that every natural number n has the property Q.

Prove that this form of the induction principle is equivalent to the one discussed in the text.

2.2 Use induction to derive the fact that the sum of the squares of the first n natural numbers is equal to

$$
\frac{2n^3 + 3n^2 + n}{6}.
$$

2.3 Use induction to establish a formula for the sum of the cubes of the first n natural numbers.

2.4 Use induction to show that if S is a set with N elements then the number of subsets of S is 2^N. (Hint: don't forget the empty set!)

2.5 Use induction to show that the sum of the first m positive even integers is equal to $m \cdot (m + 1)$.

2.6 Consider finitely many circles in the plane, possibly of different radii, and intersecting each other. These curves separate the plane into finitely many different regions.

   Prove, using induction, that these regions can always be colored red, blue, or yellow, so that no two regions sharing a nontrivial common boundary curve will be the same color.

2.7 Let $S = \{a, b, c\}$. List all possible equivalence relations on the set S.

2.8 The Well Ordering Principle, as applied to the natural numbers $\mathbb{N}$, says the following:

   If S is a nonempty subset of $\mathbb{N}$ then S has a least element.

   Here $s \in S$ is said to be a least element if for any $x \in S$ it holds that $s \leq x$.

   Assume that the natural numbers satisfy the Well Ordering Principle (this is in fact true, but further explanation requires more set theory and logic than we can cover here). If $S \subseteq \mathbb{N}$, then prove that the least element of S is unique.

   Show that the Well Ordering Principle *implies* the Induction Principle. (Hint: Assume the hypotheses of the Induction Principle. If it is not the case that $Q(x)$ is true for all $x \in \mathbb{N}$ then let S be the set of x for which $Q(x)$ is false. By Well Ordering, S has a least element s. This leads to a contradiction.)

2.9 Here is an old problem that can be found in many puzzle books: You are given nine pearls. All of these pearls except one have the same weight. Using just a balance scale, find the odd pearl in just three weighings.

   You might try your hand at this for fun. Now here is a bogus proof that you can find the odd pearl among *any finite number* of pearls in just three weighings:
   - If there are $n = 1$ pearls then the problem is trivial.
   - Assume that the problem has been solved for n pearls.
   - To solve the problem for $n + 1$ pearls, remove one pearl and put it in your pocket. Since you have solved the problem for n pearls, you can apply this solution to the remaining n pearls. If it works and you find the odd pearl, you are done. If not, the odd pearl is the one that you placed in your pocket.

   What is wrong with this reasoning? (Hint: the error here is quite different from the one in the third example in the text.)

2.10 Let f be a function with domain the reals and range the reals. Assume that f has a local minimum at each point x in its domain. (This means that for each $x \in \mathbb{R}$ there is an $\epsilon > 0$ such that whenever $|x - t| < \epsilon$ then $f(x) \leq f(t)$). *Do not assume that f is differentiable, or continuous, or anything nice like that.* Prove that the image of f is countable. (Hint: When I solved this problem as a student my solution was ten pages long; however there is a one line solution due to Michael Spivak.)

2.11 Let S be the set of all living people. Tell which of the following are equivalence relations on S. Give detailed reasons for your answers.

x is related to y if x and y are siblings.

x is related to y if y is presently a spouse of x.

x is related to y if y has at one time or another been a spouse of x.

x is related to y if y is a parent of x.

x is related to y if y is a child of x.

2.12 Let S be the set of all integers. Say that x is related to y if 3 divides $y - x$. Is this an equivalence relation on S? What if 3 is replaced by some other nonzero integer n?

2.13 Let S be the collection of all polynomials with real coefficients. Say that p is related to q if the number 0 is a root of $p - q$. Is this an equivalence relation on S?

2.14 Let $\mathcal{S}$ be the set of all subsets of the real numbers. Say that $X \in \mathcal{S}$ is related to $Y \in \mathcal{S}$ if $card(X) = card(Y)$. Is this an equivalence relation on $\mathcal{S}$?

2.15 Let S be the set of all pairs of real numbers (x, y) with $y \neq 0$. Declare two pairs (x, y) and (x', y') to be related if $x \cdot y' = x' \cdot y$. Let the set of all equivalence classes be called **R**. Emulating the construction of the rational numbers, define notions of addition and multiplication on **R**. Set up a natural bijection between $\mathbb{R}$ and **R** that respects the operations of multiplication and addition.

What conclusion do you draw from this exercise?

2.16 Perform Exercise 15 with $\mathbb{R}$ replaced by the complex numbers.

2.17 Imitate the proof of the unambiguity of addition in the integers to establish the unambiguity of subtraction and multiplication.

2.18 Let $x = [(a, b)]$ be an integer. Define $|x|$ to be $b - a$ if $b > a$, $a - b$ if $a > b$, and 0 otherwise. Prove that this definition is unambiguous.

Prove that if x and y are integers and $|x| > |y|$ then there is no nonzero integer z such that $x \cdot z = y$.

2.19 Take the commutativity, associativity, and distributivity of addition and multiplication in the natural number system for granted. That is, if $x, y, z \in \mathbb{N}$ then $x + y = y + x, x \cdot y = y \cdot x, x + (y + z) = (x + y) + z, x \cdot (y \cdot z) = (x \cdot y) \cdot z, x \cdot (y + z) = x \cdot y + x \cdot z$. Prove corresponding properties for addition and multiplication of integers.

2.20 Prove that addition of rational numbers is unambiguous.

2.21 Prove the parts of Theorem 2.8 that were not proved in the text.

2.22 Prove parts (2) and (4) of Theorem 2.9.

2.23 Let A be a set of real numbers that is bounded above and set $\alpha = \sup A$. Let $B = \{-a : a \in A\}$. Prove that $\inf B = -\alpha$. Prove the same result with the roles of infimum and supremum reversed.

2.24 Taking the commutative, associative, and distributive laws for the real number system for granted, establish these laws for the complex numbers.

2.25 Consider the function $\phi : \mathbb{R} \to \mathbb{C}$ given by $\phi(x) = x + i \cdot 0$. Prove that ϕ respects addition and multiplication in the sense that $\phi(x + x') = \phi(x) + \phi(x')$ and $\phi(x \cdot x') = \phi(x) \cdot \phi(x')$.

2.26 If $z, w \in \mathbb{C}$ and $w \neq 0$ then prove that $\overline{z/w} = \overline{z}/\overline{w}$.

2.27 Complete the calculation in the proof of Theorem 2.12.

2.28 Prove that the set of all complex numbers is uncountable.

2.29 Prove that the set of all complex numbers with rational real part is uncountable.

2.30 Prove that the set of all complex numbers with both real and imaginary parts rational is countable.

2.31 Prove that the set $\{z \in \mathbb{C} : |z| = 1\}$ is uncountable.

2.32 Prove that the field of complex numbers cannot be made into an *ordered* field. (Hint: Since $i \neq 0$ then either $i > 0$ or $i < 0$. Both lead to a contradiction.)

2.33 Let λ be a positive irrational real number. If n is a positive integer, choose by the Archimedean Property an integer k such that $k\lambda \leq n < (k+1)\lambda$. Let $\varphi(n) = n - k\lambda$. Prove that the set of all $\varphi(n)$ is dense in the interval $[0, \lambda]$. (Hint: Examine the proof of the density of the rationals in the reals.)

2.34 Prove the last statement of Section 5 without using results from later in the chapter.

2.35 Let n be a natural number and x a positive real number. Prove that there is a positive real number y such that $y^n = x$. Is y unique?

2.36 Prove that if n is a positive integer that is the square of a rational number then in fact it is the square of an integer.

Appendix: Construction of the Real Numbers

There are several techniques for constructing the real number system $\mathbb{R}$ from the rational numbers system $\mathbb{Q}$. We use the method of Dedekind (Julius W. R. Dedekind, 1831–1916) cuts because it uses a minimum of new ideas and is fairly brief.

DEFINITION 2.26   *A **cut** is a subset C of $\mathbb{Q}$ with the following properties:*

- *$C \neq \emptyset$.*
- *If $s \in C$ and $t < s$ then $t \in C$.*
- *If $s \in C$ then there is a $u \in C$ such that $u > s$.*
- *There is a rational number x such that $c < x$ for all $c \in C$.*

You should think of a cut C as the set of all rational numbers to the left of some point in the real line. Since we have not constructed the real line yet, we cannot define a cut in that simple way; we have to make the construction more indirect. But if you consider the three properties of a cut, they describe a set that looks like a "rational left half-line."

Notice that if C is a cut and $s \notin C$ then any rational $t > s$ is also not in C. Also, if $r \in C$ and $s \notin C$ then it must be that $s > r$.

DEFINITION 2.27   *If C and D are cuts then we say that $C < D$, provided that C is a subset of D but $C \neq D$.*

Check for yourself that "$<$" is an ordering on the set of all cuts.

Now we introduce operations of addition and multiplication which will turn the set of all cuts into a field.

DEFINITION 2.28   *If C and D are cuts then we define*

$$C + D = \{c + d : c \in C, d \in D\}.$$

We define the cut $\hat{0}$ to be the set of all negative rationals.

The cut $\hat{0}$ will play the role of the additive identity. We are now required to check that field axioms A1–A5 hold.

For A1, we need to see that $C + D$ is a cut. Obviously $C + D$ is not empty. If s is an element of $C + D$ and t is a rational number less than s, write $s = c + d$, where $c \in C$ and $d \in D$. Then $t - c < s - c = d \in D$ so $t - c \in D$; and $c \in C$. Hence $t = c + (t - c) \in C + D$. A similar argument shows that there is an $r > s$ such that $r \in C + D$. Finally, if x is a rational upper bound for C and y

is a rational upper bound for $\mathcal{D}$, then $x + y$ is a rational upper bound for $\mathcal{C} + \mathcal{D}$. We conclude that $\mathcal{C} + \mathcal{D}$ is a cut.

Since addition of rational numbers is commutative, it follows immediately that addition of cuts is commutative. Associativity follows in a similar fashion.

Now we show that if $\mathcal{C}$ is a cut then $\mathcal{C} + \hat{0} = \mathcal{C}$. For if $c \in \mathcal{C}$ and $z \in \hat{0}$ then $c + z < c + 0 = c$ hence $\mathcal{C} + \hat{0} \subseteq \mathcal{C}$. Also, if $c' \in \mathcal{C}$ then choose a $d' \in \mathcal{C}$ such that $c' < d'$. Then $c' - d' < 0$ so $c' - d' \in \hat{0}$. And $c' = d' + (c' - d')$. Hence $\mathcal{C} \subseteq \mathcal{C} + \hat{0}$. We conclude that $\mathcal{C} + \hat{0} = \mathcal{C}$.

Finally, for Axiom A5, we let $\mathcal{C}$ be a cut and set $-\mathcal{C}$ to be equal to $\{d \in \mathbb{Q} : c + d < 0$ for all $c \in \mathcal{C}\}$. If x is a rational upper bound for $\mathcal{C}$ and $c \in \mathcal{C}$ then $-x \in -\mathcal{C}$ so $-\mathcal{C}$ is not empty. By its very definition, $\mathcal{C} + (-\mathcal{C}) \subseteq \hat{0}$. Further, if $z \in \hat{0}$ and $c \in \mathcal{C}$ we set $c' = z - c$. Then $c' \in -\mathcal{C}$ and $z = c + c'$. Hence $\hat{0} \subseteq \mathcal{C} + (-\mathcal{C})$. We conclude that $\mathcal{C} + (-\mathcal{C}) = \hat{0}$.

Having verified the axioms for addition, we turn now to multiplication.

DEFINITION 2.29 *If $\mathcal{C}$ and $\mathcal{D}$ are cuts then we define the product $\mathcal{C} \cdot \mathcal{D}$ as follows:*

> *If $\mathcal{C}, \mathcal{D} > \hat{0}$ then $\mathcal{C} \cdot \mathcal{D} = \{q \in \mathbb{Q} : q < c \cdot d$ for some $c \in \mathcal{C}, d \in \mathcal{D}$ with $c > 0, d > 0\}$.*
> *If $\mathcal{C} > \hat{0}, \mathcal{D} < \hat{0}$ then $\mathcal{C} \cdot \mathcal{D} = -(\mathcal{C} \cdot (-\mathcal{D}))$.*
> *If $\mathcal{C} < \hat{0}, \mathcal{D} > \hat{0}$ then $\mathcal{C} \cdot \mathcal{D} = -((-\mathcal{C}) \cdot \mathcal{D})$.*
> *If $\mathcal{C}, \mathcal{D} < \hat{0}$ then $\mathcal{C} \cdot \mathcal{D} = (-\mathcal{C}) \cdot (-\mathcal{D})$.*
> *If either $\mathcal{C} = \hat{0}$ or $\mathcal{D} = \hat{0}$ then $\mathcal{C} \cdot \mathcal{D} = \hat{0}$.*

Notice that, for convenience, we have defined multiplication of negative numbers just as we did in high school. The reason is that the definition that we use for the product of two positive numbers cannot work when one of the two factors is negative (exercise).

It is now a routine exercise to verify that the set of all cuts, with this definition of multiplication, satisfies field axioms M1–M5. The proofs follow those for A1–A5 rather closely.

For the distributive property, one first checks the case when all the cuts are positive, reducing it to the distributive property for the rationals. Then one handles negative cuts on a case-by-case basis.

We now know that the collection of all cuts forms an ordered field. Denote this field by the symbol $\mathbb{R}$. We next verify the crucial property of $\mathbb{R}$ that sets it apart from $\mathbb{Q}$:

THEOREM 2.20
The ordered field $\mathbb{R}$ satisfies the least upper bound property.

PROOF   Let S be a subset of $\mathbb{R}$ that is bounded above. Define

$$S^* = \bigcup_{\mathcal{C} \in S} \mathcal{C}.$$

Then S^* is clearly nonempty, and it is therefore a cut since it is a union of cuts. It is also clearly an upper bound for S since it contains each element of S. It remains to check that S^* is the least upper bound for S.

In fact if $\mathcal{T} < S^*$ then $\mathcal{T} \subseteq S^*$ and there is a rational number q in $S^* \setminus \mathcal{T}$. But, by the definition of S^*, it must be that $q \in \mathcal{C}$ for some $\mathcal{C} \in S$. So $\mathcal{C} > \mathcal{T}$, and $\mathcal{T}$ cannot be an upper bound for S. Therefore S^* is the least upper bound for S, as desired.   ∎

We have shown that $\mathbb{R}$ is an ordered field that satisfies the least upper bound property. It remains to show that $\mathbb{R}$ contains (a copy of) $\mathbb{Q}$ in a natural way. In fact, if $q \in \mathbb{Q}$ we associate to it the element $\varphi(q) = \mathcal{C}_q \equiv \{x \in \mathbb{Q} : x < q\}$. Then $\mathcal{C}_q$ is obviously a cut. It is also routine to check that

$$\varphi(q + q') = \varphi(q) + \varphi(q') \quad \text{and} \quad \varphi(q \cdot q') = \varphi(q) \cdot \varphi(q').$$

Therefore we see that ϕ represents $\mathbb{Q}$ as a subfield of $\mathbb{R}$.

3

Sequences

3.1 Convergence of Sequences

A *sequence* of real numbers is a function $\varphi : \mathbb{N} \to \mathbb{R}$. We often write the sequence as $\varphi(1), \varphi(2), \ldots$ or, more simply, as $\varphi_1, \varphi_2, \ldots$. A sequence of complex numbers is defined similarly, with $\mathbb{R}$ replaced by $\mathbb{C}$.

Example 3.1

The function $\varphi(j) = 1/j$ is a sequence of real numbers. We will often write such a sequence as $\{1, 1/2, 1/3, \ldots\}$ or as $\{1/j\}_{j=1}^{\infty}$.

The function $\psi(j) = \cos j + i \sin j$ is a sequence of complex numbers.

Do not be misled into thinking that a sequence must form a pattern or be given by a formula. Obviously the ones which are given by a formula are easy to write down, but they are certainly not typical. For example, the coefficients in the decimal expansion of π, $\{3, 1, 4, 1, 5, 9, 2, 6, 5, \ldots\}$, fit our definition of sequence, but they are not given by any obvious pattern. ⬚

The most important question about a sequence is whether it converges. We define this notion as follows.

DEFINITION 3.2 *A sequence $\{a_j\}$ of real (resp. complex) numbers is said to **converge** to a real (resp. complex) number α if for each $\epsilon > 0$ there is an integer $N > 0$ such that if $j > N$ then $|a_j - \alpha| < \epsilon$. We call α the **limit** of the sequence $\{a_j\}$. We sometimes write $a_j \to \alpha$.*

If a sequence $\{a_j\}$ does not converge then we frequently say that it *diverges.*

Example 3.3

Let $a_j = 1/j, j = 1, 2, \dots$. Then the sequence converges to 0. For let $\epsilon > 0$. Choose N to be the next integer after $1/\epsilon$ (we use here the Archimedean principle). If $j > N$ then

$$|a_j - 0| = |a_j| = \frac{1}{j} < \frac{1}{N} < \epsilon,$$

proving the claim.

Let $b_j = (-1)^j, j = 1, 2, \dots$. Then the sequence *does not converge*. To prove this assertion, suppose that it does. Say that the sequence converges to a number α. Let $\epsilon = 1/2$. By definition of convergence, there is an integer $N > 0$ such that if $j > N$ then $|b_j - \alpha| < \epsilon = 1/2$. For such j we have

$$|b_j - b_{j+1}| \le |b_j - \alpha| + |\alpha - b_{j+1}|$$

(by the triangle inequality — Proposition 2.18). But this last is

$$< \epsilon + \epsilon = 1.$$

On the other hand,

$$|b_j - b_{j+1}| = \left|(-1)^j - (-1)^{j+1}\right| = 2.$$

The last two lines yield that $2 < 1$, a clear contradiction. So the sequence $\{b_j\}$ has no limit. □

We begin with a few intuitively appealing properties of convergent sequences which will be needed later. First, a definition.

DEFINITION 3.4   *A sequence a_j is said to be **bounded** if there is a number $M > 0$ such that $|a_j| \le M$ for every j.*

Now we have

PROPOSITION 3.1

Let $\{a_j\}$ be a convergent sequence. Then we have

- *The limit of the sequence is unique.*
- *The sequence is bounded.*

PROOF   Suppose that the sequence has two limits α and α'. Let $\epsilon > 0$. Then there is an integer $N > 0$ such that for $j > N$ we have $|a_j - \alpha| < \epsilon$. Likewise, there is an integer $N' > 0$ such that for $j > N'$ we have $|a_j - \alpha'| < \epsilon$. Let

$N_0 = \max\{N, N'\}$. Then for $j > N_0$ we have

$$|\alpha - \alpha'| \leq |\alpha - a_j| + |a_j - \alpha'| < \epsilon + \epsilon = 2\epsilon.$$

Since this inequality holds for any $\epsilon > 0$ we have that $\alpha = \alpha'$.

Next, with α the limit of the sequence and $\epsilon = 1$ we choose an integer $N > 0$ such that $j > N$ implies that $|a_j - \alpha| < \epsilon = 1$. For such j we have that

$$|a_j| \leq |a_j - \alpha| + |\alpha| < 1 + |\alpha| \equiv P.$$

Let $Q = \max\{|a_1|, |a_2|, \ldots, |a_N|\}$. If j is any natural number then either $1 \leq j \leq N$, in which case $|a_j| \leq Q$, or else $j > N$, in which case $|a_j| \leq P$. Set $M = \max\{P, Q\}$. Then $|a_j| \leq M$ for all j, as desired. So the sequence is bounded. ∎

The next proposition records some elementary properties of limits of sequences.

PROPOSITION 3.2

Let $\{a_j\}$ be a sequence of real or complex numbers with limit α and $\{b_j\}$ be a sequence of real or complex numbers with limit β. Then we have

1. *If c is a constant then the sequence $\{c \cdot a_j\}$ converges to $c \cdot \alpha$.*
2. *The sequence $\{a_j + b_j\}$ converges to $\alpha + \beta$.*
3. *The sequence $a_j \cdot b_j$ converges to $\alpha \cdot \beta$.*
4. *If $b_j \neq 0$ for all j and $\beta \neq 0$ then the sequence a_j / b_j converges to α / β.*

PROOF   For the first part, we may assume that $c \neq 0$ (for when $c = 0$ there is nothing to prove). Let $\epsilon > 0$. Choose an integer $N > 0$ such that for $j > N$ it holds that

$$|a_j - \alpha| < \frac{\epsilon}{|c|}.$$

For such j we have that

$$|c \cdot a_j - c \cdot \alpha| = |c| \cdot |a_j - \alpha| < |c| \cdot \frac{\epsilon}{|c|} = \epsilon.$$

This proves the first assertion. The proof of the second assertion is similar, and we leave it as an exercise. For the third assertion, notice that the sequence $\{a_j\}$ is bounded: say that $|a_j| \leq M$ for every j. Let $\epsilon > 0$. Choose an integer $N > 0$

so that $|a_j - \alpha| < \epsilon/(2M + 2|\beta|)$ when $j > N$. Also choose an integer $N' > 0$ such that $|b_j - \beta| < \epsilon/(2M + 2|\beta|)$ when $j > N'$. Then for $j > \max\{N, N'\}$ we have that

$$
\begin{aligned}
|a_j b_j - \alpha\beta| &= |a_j(b_j - \beta) + \beta(a_j - \alpha)| \\
&\leq |a_j(b_j - \beta)| + |\beta(a_j - \alpha)| \\
&< M \cdot \frac{\epsilon}{2M + 2|\beta|} + |\beta| \cdot \frac{\epsilon}{2M + 2|\beta|} \\
&\leq \frac{\epsilon}{2} + \frac{\epsilon}{2} \\
&= \epsilon.
\end{aligned}
$$

So the sequence $\{a_j b_j\}$ converges to $\alpha\beta$.

Part (4) is proved in a similar fashion and we leave the details as an exercise.∎

REMARK 3.3   You were probably puzzled by the choice of N and N' in the proof of part (3) of Proposition 3.2. Where did the number $\epsilon/(2M + 2|\beta|)$ come from? The answer of course becomes obvious when we read on further in the proof. So the lesson here is that a proof is constructed backward: you look to the end of the proof to see what you need to specify earlier on.   ∎

When discussing the convergence of a sequence, we often find it inconvenient to deal with the definition of convergence as given. For this definition makes reference to the number to which the sequence is supposed to converge, and we often do not know this number in advance. Wouldn't it be useful to be able to decide whether a sequence converges *without knowing to what it converges*?

DEFINITION 3.5   *Let $\{a_j\}$ be a sequence of real (resp. complex) numbers. We say that the sequence satisfies the **Cauchy criterion** (A. L. Cauchy, 1789–1857) — more briefly, that the sequence is **Cauchy** — if for each $\epsilon > 0$ there is an integer $N > 0$ such that if $j, k > N$ then $|a_j - a_k| < \epsilon$.*

Notice that the concept of a sequence being Cauchy simply makes precise the notion of the elements of the sequence (i) *getting* close together and (ii) *staying* close together.

LEMMA 3.4
Every Cauchy sequence is bounded.

PROOF   Let $\epsilon = 1 > 0$. There is an integer $N > 0$ such that $|a_j - a_k| < \epsilon = 1$

whenever $j, k > N$. Thus if $j \geq N + 1$ we have

$$|a_j| \leq |a_{N+1} + (a_j - a_{N+1})|$$
$$\leq |a_{N+1}| + |a_j - a_{N+1}|$$
$$\leq |a_{N+1}| + 1 \equiv K.$$

Let $L = \max\{|a_1|, |a_2|, \ldots, |a_N|\}$. If j is any natural number, then either $1 \leq j \leq N$, in which case $|a_j| \leq L$, or $j > N$, in which case $|a_j| \leq K$. Set M $= \max\{K, L\}$. Then, for any j, $|a_j| \leq M$ as required.   ∎

THEOREM 3.5

Let $\{a_j\}$ be a sequence of real numbers. The sequence is Cauchy if and only if it converges to some limit α.

PROOF   First assume that the sequence converges to a limit α. Let $\epsilon > 0$. Choose, by definition of convergence, an integer $N > 0$ such that if $j > N$ then $|a_j - \alpha| < \epsilon/2$. If $j, k > N$ then

$$|a_j - a_k| \leq |a_j - \alpha| + |\alpha - a_k| < \frac{\epsilon}{2} + \frac{\epsilon}{2} = \epsilon.$$

So the sequence is Cauchy.

Conversely, suppose that the sequence is Cauchy. Define

$$S = \{x \in \mathbb{R} : x < a_j \text{ for all but finitely many } j\}.$$

By the lemma, the sequence $\{a_j\}$ is bounded by some number M. If x is a real number less than $-M$ then $x \in S$, so S is nonempty. Also, S is bounded above by M. Let $\alpha = \sup S$. Then α is a well-defined real number, and we claim that α is the limit of the sequence $\{a_j\}$.

To see this, let $\epsilon > 0$. Choose an integer $N > 0$ such that $|a_j - a_k| < \epsilon/2$ whenever $j, k > N$. Notice that this last inequality implies that

$$|a_j - a_{N+1}| < \epsilon/2 \text{ when } j \geq N + 1 \tag{$*$}$$

hence

$$a_j > a_{N+1} - \epsilon/2 \text{ when } j \geq N + 1.$$

Thus $a_{N+1} - \epsilon/2 \in S$ and it follows that

$$\alpha \geq a_{N+1} - \epsilon/2. \tag{$**$}$$

Line $(*)$ also shows that

$$a_j < a_{N+1} + \epsilon/2 \text{ when } j \geq N + 1.$$

Thus $a_{N+1} + \epsilon/2 \notin S$ and

$$\alpha \le a_{N+1} + \epsilon/2. \qquad (***)$$

Combining lines $(**)$ and $(***)$ gives

$$|\alpha - a_{N+1}| \le \epsilon/2.$$

But then line $(*)$ yields, for $j > N$, that

$$|\alpha - a_j| \le |\alpha - a_{N+1}| + |a_{N+1} - a_j| < \epsilon/2 + \epsilon/2 = \epsilon.$$

This proves that the sequence $\{a_j\}$ converges to α, as claimed. ∎

COROLLARY 3.6

Let $\{\alpha_j\}$ be a sequence of complex numbers. The sequence is Cauchy if and only if it is convergent.

PROOF Write $\alpha_j = a_j + ib_j$, with a_j, b_j real. Then $\{\alpha_j\}$ is Cauchy if and only if $\{a_j\}$ and $\{b_j\}$ are Cauchy. Also $\{\alpha_j\}$ is convergent to a complex limit α if and only if $\{a_j\}$ converges to Re α and $\{b_j\}$ converges to Im α. These observations, together with the theorem, prove the corollary. ∎

DEFINITION 3.6 *Let $\{a_j\}$ be a sequence of real numbers. The sequence is said to be **monotone increasing** if $a_1 \le a_2 \le \dots$. It is **monotone decreasing** if $a_1 \ge a_2 \ge \dots$.*

The word "monotone" is used here primarily for reasons of tradition. In many contexts the word is redundant and we omit it.

PROPOSITION 3.7

If $\{b_j\}$ is a monotone increasing sequence that is bounded above — $a_j \le M$ for all j — then $\{a_j\}$ is convergent. If $\{b_j\}$ is a monotone decreasing sequence that is bounded below — $b_j \ge K > -\infty$ for all j — then $\{b_j\}$ is convergent.

PROOF Let $\epsilon > 0$. Let $\alpha = \sup a_j < \infty$. By definition of supremum there is an integer $N > 0$ such that if $j > N$ then $|a_j - \alpha| < \epsilon$. Then if $\ell \ge N + 1$ we have $a_j \le a_\ell \le \alpha$ hence $|a_\ell - \alpha| < \epsilon$. Thus the sequence converges to α.

The proof for monotonically decreasing sequences is similar and we omit it. ∎

A proof very similar to that of the proposition gives the following useful fact.

COROLLARY 3.8
Let S be a set of real numbers that is bounded above and below. Let β be its supremum and α its infimum. If $\epsilon > 0$ then there are $s, t \in S$ such that $|s - \beta| < \epsilon$ and $|t - \alpha| < \epsilon$.

PROOF   This is a restatement of the proof of the proposition.   ∎

We conclude the section by recording one of the most useful results for calculating the limit of a sequence:

PROPOSITION 3.9   THE PINCHING PRINCIPLE
Let $\{a_j\}, \{b_j\},$ and $\{c_j\}$ be sequences of real numbers satisfying

$$a_j \leq b_j \leq c_j$$

for every j. If

$$\lim_{j \to \infty} a_j = \lim_{j \to \infty} c_j = \alpha$$

for some real number α then

$$\lim_{j \to \infty} b_j = \alpha.$$

PROOF   This proof is requested of you in the Exercises.   ∎

3.2   Subsequences

Let $\{a_j\}$ be a given sequence. If

$$0 < j_1 < j_2 < \ldots$$

are positive integers then the function

$$k \mapsto a_{j_k}$$

is called a *subsequence* of the given sequence. We usually write the subsequence as

$$\{a_{j_k}\}_{k=1}^{\infty} \quad \text{or} \quad \{a_{j_k}\}.$$

Example 3.7
Consider the sequence

$$\{2^j\} = \{2, 4, 8, \ldots\}.$$

Then the sequence

$$\{2^{2k}\} = \{4, 16, 64, \ldots\}$$

is a subsequence. Notice that the subsequence contains a subcollection of elements in the original sequence *in the original order*. In this example, $j_k = 2k$.

Another subsequence is

$$\{2^{(2^k)}\} = \{4, 16, 256, \ldots\}.$$

In this instance, it holds that $j_k = 2^k$. Notice that this new subsequence is in fact a subsequence of the first subsequence. That is, it is a sub-subsequence of the original sequence $\{2^j\}$.   ⬚

PROPOSITION 3.10

If $\{a_j\}$ is a convergent sequence with limit α, then every subsequence converges to the limit α.

Conversely, if a sequence $\{b_j\}$ has the property that each of its subsequences is convergent then $\{b_j\}$ itself is convergent.

PROOF   Assume $\{a_j\}$ is convergent to a limit α, and let $\{a_{j_k}\}$ be a subsequence. Let $\epsilon > 0$ and choose $N > 0$ such that $|a_j - \alpha| < \epsilon$ whenever $j > N$. Now if $k > N$ then $j_k > N$ hence $|a_{j_k} - \alpha| < \epsilon$. Therefore, by definition, the subsequence $\{a_{j_k}\}$ also converges to α.

The converse is trivial, simply because the sequence is a subsequence of itself.   ∎

See Exercise 7 for a powerful generalization of the converse of this proposition.

Now we present one of the most fundamental theorems of basic real analysis (due to B. Bolzano, 1781–1848, and K. Weierstrass, 1815–1897).

THEOREM 3.11   BOLZANO–WEIERSTRASS

Let $\{a_j\}$ be a bounded sequence in $\mathbb{R}$. Then there is a subsequence that converges.

PROOF   Say that $|a_j| \leq M$ for every j. We may assume that $M \neq 0$.

One of the two intervals $[-M, 0]$ and $[0, M]$ must contain infinitely many elements of the sequence. Say that $[0, M]$ does. Choose a_{j_1} to be one of the infinitely many sequence elements in $[0, M]$.

Next, one of the intervals $[0, M/2]$ and $[M/2, M]$ must contain infinitely many elements of the sequence. Say that it is $[0, M/2]$. Choose an element a_{j_2}, with $j_2 > j_1$, from $[0, M/2]$. Continue in this fashion, halving the interval, choosing a half with infinitely many sequence elements, and selecting the next subsequence element from that half.

Let us analyze the resulting subsequence. Notice that $|a_{j_1} - a_{j_2}| \leq M$ since both elements belong to the interval $[0, M]$. Likewise, $|a_{j_2} - a_{j_3}| \leq M/2$ since both elements belong to $[0, M/2]$. In general, $|a_{j_k} - a_{j_{k+1}}| \leq 2^{-k+1} \cdot M$ for each $k \in \mathbb{N}$. Now let $\epsilon > 0$. Choose an integer $N > 0$ such that $2^{-N} < \epsilon/(2M)$. Then for any $m > l > N$ we have

$$
\begin{aligned}
|a_{j_l} - a_{j_m}| &= |(a_{j_l} - a_{j_{l+1}}) + (a_{j_{l+1}} - a_{j_{l+2}}) + \cdots + (a_{j_{m-1}} - a_{j_m})| \\
&\leq |a_{j_l} - a_{j_{l+1}}| + |a_{j_{l+1}} - a_{j_{l+2}}| + \cdots + |a_{j_{m-1}} - a_{j_m}| \\
&\leq 2^{-l+1} \cdot M + 2^{-l} \cdot M \cdots \\
&\quad + 2^{-m+2} \cdot M \\
&= \left(2^{-l+1} + 2^{-l} + 2^{-l-1} + \cdots + 2^{-m+2}\right) \cdot M \\
&= \left((2^{-l+2} - 2^{-l+1}) + (2^{-l+1} - 2^{-l}) + \cdots \right. \\
&\quad \left. + (2^{-m+3} - 2^{-m+2})\right) \cdot M \\
&= \left(2^{-l+2} - 2^{-m+2}\right) \cdot M \\
&< 2^{-l+2} \cdot M \\
&< 2 \cdot \frac{\epsilon}{2M} \cdot M \\
&= \epsilon.
\end{aligned}
$$

We see that the subsequence $\{a_{j_k}\}$ is Cauchy, so it converges. ∎

REMARK 3.12   The Bolzano–Weierstrass Theorem is a generalization of our result from the last section about monotone increasing sequences that are bounded above (resp. monotone decreasing sequences that are bounded below). For such a sequence is surely bounded above *and* below (why?). So it has a convergent subsequence. And thus it follows easily that the entire sequence converges. Details are left as an exercise. ∎

Example 3.8

In this text we have not yet given a rigorous definition of the function $\sin x$. However, just for the moment, use the definition you learned in calculus class and consider the sequence $\{\sin j\}_{j=1}^{\infty}$. Notice that the sequence is bounded in absolute value by 1. The Bolzano–Weierstrass theorem guarantees that there is a convergent subsequence, even though it would be very difficult to say what that convergent subsequence is. ☐

COROLLARY 3.13

Let $\{\alpha_j\}$ *be a bounded sequence of* **complex** *numbers. Then there is a convergent subsequence.*

PROOF Write $\alpha_j = a_j + ib_j$, with $a_j, b_j \in \mathbb{R}$. The fact that $\{\alpha_j\}$ is bounded implies that $\{a_j\}$ is bounded. By the Bolzano–Weierstrass theorem, there is a convergent subsequence $\{a_{j_k}\}$.

Now the sequence $\{b_{j_k}\}$ is bounded. So it has a convergent subsequence $\{b_{j_{k_l}}\}$. Then the sequence $\{\alpha_{j_{k_l}}\}$ is convergent, and is a subsequence of the original sequence $\{\alpha_j\}$. ∎

In earlier parts of this chapter we have discussed sequences that converge to a finite number. Such a sequence is, by Proposition 3.1, bounded. However, in some mathematical contexts, it is useful to speak of a sequence "converging to infinity." Obviously this notion of convergence is separate and distinct from the notion that we have been discussing until now. Context always makes clear which type of convergence is meant. Now we will briefly treat the idea of "convergence to infinity."

DEFINITION 3.9 *We say that a sequence* $\{a_j\}$ *of real numbers converges to* $+\infty$ *if for every* $M > 0$ *there is an integer* $N > 0$ *such that* $a_j > M$ *whenever* $j > N$. *We write* $a_j \to +\infty$. *We say that* $\{a_j\}$ *converges to* $-\infty$ *if for every* $M > 0$ *there is an integer* $N > 0$ *such that* $a_j < -M$ *whenever* $j > N$. *We write* $a_j \to -\infty$.

REMARK 3.14 Notice that the statement $a_j \to +\infty$ means that we can make a_j become arbitrarily large and positive and *stay* large and positive just by making j large enough.

Likewise, the statement $a_j \to -\infty$ means that we can force a_j to be arbitrarily large and negative, and *stay* large and negative, just by making j large enough. ∎

Example 3.10

The sequence $\{j^2\}$ converges to $+\infty$. The sequence $\{-2j + 18\}$ converges to $-\infty$. The sequence $\{j + (-1)^j \cdot j\}$ has no infinite limit and no finite limit. However the subsequence $\{0, 0, 0, \ldots\}$ converges to 0 and the subsequence $\{4, 8, 12 \ldots\}$ converges to $+\infty$. You are asked to supply details in Exercise 8. ☐

With the new language provided by Definition 3.9, we may generalize Proposition 3.7.

PROPOSITION 3.15

Let $\{a_j\}$ be a monotone increasing sequence of real numbers. Then the sequence has a limit — either a finite number or $+\infty$.

Let $\{b_j\}$ be a monotone decreasing sequence of real numbers. Then the sequence has a limit — either a finite number or $-\infty$.

In the same spirit as the last definition, we also have the following:

DEFINITION 3.11   *If S is a nonempty set of real numbers that is **not** bounded above, we say that its supremum (or least upper bound) is $+\infty$.*

*If T is a nonempty set of real numbers which is **not** bounded below then we say that its infimum (or greatest lower bound) is $-\infty$.*

Exercise 9 asks you to explain why logic forces us to declare the supremum of the empty set to be $-\infty$ and the infimum of the empty set to be $+\infty$.

3.3   Lim sup and Lim inf

Convergent sequences are useful objects, but the unfortunate truth is that most sequences do not converge. Nevertheless, we would like to have a language for discussing the asymptotic behavior of any real sequence $\{a_j\}$ as $j \to \infty$. That is the purpose of the concepts of "limit superior" (or "upper limit") and "limit inferior" (or "lower limit").

DEFINITION 3.12   *Let $\{a_j\}$ be a sequence of real numbers. For each j let*

$$A_j = \inf\{a_j, a_{j+1}, a_{j+2}, \ldots\}.$$

Then $\{A_j\}$ is a monotone increasing sequence (since as j becomes large we are taking the infimum of a smaller set of numbers), so it has a limit. We define the limit inferior of $\{a_j\}$ to be

$$\liminf_{j \to \infty} a_j = \lim_{j \to \infty} A_j.$$

Likewise, let

$$B_j = \sup\{a_j, a_{j+1}, a_{j+2}, \ldots\}.$$

Then $\{B_j\}$ is a monotone decreasing sequence (since as j becomes large we are taking the supremum of a smaller set of numbers), so it has a limit. We define the limit superior of $\{a_j\}$ to be

$$\limsup_{j \to \infty} a_j = \lim_{j \to \infty} B_j.$$

REMARK 3.16 What is the intuitive content of this definition? For each j, A_j picks out the greatest lower bound of the sequence in the jth position or later. So the sequence $\{A_j\}$ should tend to the *smallest* possible limit of any subsequence of $\{a_j\}$.

Likewise, for each j, B_j picks out the least upper bound of the sequence in the jth position or later. So the sequence $\{A_j\}$ should tend to the *greatest* possible limit of any subsequence of $\{a_j\}$. We shall make this remark more precise in Proposition 3.17 below.

Notice that it is implicit in the definition that *every* real sequence has a limit supremum and a limit infimum. ∎

Example 3.13

Consider the sequence $\{(-1)^j\}$. Of course this sequence does not converge. Let us calculate its lim sup and lim inf.

Referring to the definition, we have that $A_j = -1$ for every j. So

$$\liminf(-1)^j = \lim(-1) = -1.$$

Similarly, $B_j = +1$ for every j. Therefore

$$\limsup(-1)^j = \lim(+1) = +1.$$

As we predicted in the remark, the lim inf is the least subsequential limit, and the lim sup is the greatest subsequential limit. ☐

Now let us prove the characterizing property of lim sup and lim inf to which we have been alluding.

PROPOSITION 3.17

Let $\{a_j\}$ be a sequence of real numbers. Let $\beta = \limsup_{j \to \infty} a_j$ and $\alpha = \liminf_{j \to \infty} a_j$. If $\{a_{j_k}\}$ is any subsequence of the given sequence then

$$\alpha \leq \liminf_{k \to \infty} a_{j_k} \leq \limsup_{k \to \infty} a_{j_k} \leq \beta.$$

Moreover, there is a subsequence $\{a_{j_l}\}$ such that

$$\lim_{l \to \infty} a_{j_l} = \alpha$$

and another sequence $\{a_{j_m}\}$ such that

$$\lim_{m \to \infty} a_{j_m} = \beta.$$

PROOF   For simplicity in this proof we assume that all lim sup's and lim inf's are finite. The case of infinite lim sup's and infinite lim inf's is left to Exercise 10.

We begin by considering the liminf. We adopt the notation of Definition 3.12. There is a $j_1 \geq 1$ such that $|A_1 - a_{j_1}| < 2^{-1}$. We choose j_1 to be as small as possible. Next we choose j_2, necessarily greater than or equal to j_1, such that j_2 is as small as possible and $|a_{j_2} - A_2| < 2^{-2}$. Continuing in this fashion, we select $a_{j_k} \geq a_{j_{k-1}}$ such that $|a_{j_k} - A_k| < 2^{-k-1}$, and so forth.

Recall that $A_k \to \alpha = \liminf_{j \to \infty} a_j$. Now fix $\epsilon > 0$. If N is an integer so large that $k > N$ implies that $|A_k - \alpha| < \epsilon/2$ and also that $2^{-N} < \epsilon/2$, then for such k we have

$$|a_{j_k} - \alpha| \leq |a_{j_k} - A_k| + |A_k - \alpha|$$
$$< 2^{-k} + \frac{\epsilon}{2}$$
$$< \frac{\epsilon}{2} + \frac{\epsilon}{2}$$
$$= \epsilon.$$

Thus the subsequence $\{a_{j_k}\}$ converges to α, the lim inf of the given sequence. A similar construction gives a (different) subsequence converging to β, the lim sup of the given sequence.

Now let $\{a_{j_l}\}$ be *any* subsequence of the sequence $\{a_j\}$. Let β^* be the lim sup of this subsequence. Then, by the first part of the proof, there is a subsequence $\{a_{j_{l_m}}\}$ such that

$$\lim_{m \to \infty} a_{j_{l_m}} = \beta^*.$$

But $a_{j_{l_m}} \leq B_{j_{l_m}}$ by the very definition of the $B's$. Thus

$$\beta^* = \lim_{m \to \infty} a_{j_{l_m}} \leq \lim_{m \to \infty} B_{j_{l_m}} = \beta$$

or

$$\limsup_{l \to \infty} a_{j_l} \leq \beta,$$

as claimed. A similar argument shows that

$$\liminf_{l \to \infty} a_{j_l} \geq \alpha.$$

This completes the proof of the proposition.   ∎

COROLLARY 3.18

If $\{a_j\}$ is a sequence and $\{a_{j_k}\}$ is a convergent subsequence then

$$\liminf_{j \to \infty} a_j \leq \lim_{k \to \infty} a_{j_k} \leq \limsup_{j \to \infty} a_j.$$

Take it for granted for the moment that π has been rigorously defined and proved to be irrational (in fact we will do this in complete detail later). Then Exercise 33 of Chapter 2 shows that the positive integers are dense, modulo multiples of π, in the interval $[0, \pi]$. It follows that the sequence $\{\cos j\}$ is dense in the interval $[-1, 1]$ in the following sense: given any number $\alpha \in [-1, 1]$ there is a subsequence $\cos j_k$ such that $\lim_{k \to \infty} \cos j_k = \alpha$. In particular, the lim sup of the sequence is 1 and the lim inf is -1. You are asked to provide the details of these assertions in Exercise 11.

We close this section with a fact that is similar to one you have seen before for the supremum and infimum (Corollary 3.8). Its proof is left as Exercise 12.

PROPOSITION 3.19

Let $\{a_j\}$ be a sequence and set $\limsup a_j = \beta$ and $\liminf a_j = \alpha$. Assume that α, β are finite real numbers. Let $\epsilon > 0$. Then there are arbitrarily large j such that $a_j > \beta - \epsilon$. Also there are arbitrarily large k such that $a_k < \alpha + \epsilon$.

3.4   Some Special Sequences

We often obtain information about a new sequence by comparison with a sequence that we already know. Thus it is well to have a catalogue of fundamental sequences that provide a basis for comparison.

Example 3.14

Fix a real number a. The sequence $\{a^j\}$ is called a *power sequence*. If $-1 < a < 1$ then the sequence converges to 0. If $a = 1$ then the sequence is a constant sequence and converges to 1. If $a > 1$ then the sequence converges to ∞. Finally, if $a \leq -1$ then the sequence diverges.   ▯

Recall that in Section 2.5 we discussed the existence of nth roots of positive real numbers. If $\alpha > 0, m \in \mathbb{Z}$, and $n \in \mathbb{N}$ then we may define

$$\alpha^{m/n} = (\alpha^m)^{1/n}.$$

Thus we may talk about rational powers of a positive number. Next, if $\beta \in \mathbb{R}$ then we may define

$$\alpha^\beta = \sup\{\alpha^q : q \in \mathbb{Q}, q < \beta\}.$$

Thus we can define *any real power* of a positive real number. Exercise 13 asks you to verify several basic properties of these exponentials.

LEMMA 3.20
If $\alpha > 1$ is a real number and $\beta > 0$ then $\alpha^\beta > 1$.

PROOF Let q be a positive rational number that is less than β. Say that $q = m/n$, with m, n positive integers. It is obvious that $\alpha^m > 1$ and hence that $(\alpha^m)^{1/n} > 1$. Since α^β majorizes this last quantity, we are done. ∎

Example 3.15
Fix a real number α and consider the sequence $\{j^\alpha\}$.

If $\alpha > 0$ then it is easy to see that $j^\alpha \to +\infty$: to verify this assertion fix $M > 0$ and take the number N to be the first integer after $M^{1/\alpha}$.

If $\alpha = 0$ then j^α is a constant sequence, identically equal to 1.

If $\alpha < 0$ then $j^\alpha = 1/j^{-\alpha}$. The denominator of this last expression tends to $+\infty$, hence the sequence j^α tends to 0. □

Example 3.16
The sequence $\{j^{1/j}\}$ converges to 1. In fact, consider the expressions $\alpha_j = j^{1/j} - 1 > 0$. We have that

$$j = (\alpha_j + 1)^j \geq \frac{j(j-1)}{2}(\alpha_j)^2,$$

(the latter being just one term from the binomial expansion — see Section 2.1). Thus

$$0 < \alpha_j \leq \sqrt{2/(j-1)}$$

as long as $j \geq 2$. It follows that $\alpha_j \to 0$ or $j^{1/j} \to 1$. □

Example 3.17
Let α be a positive real number. Then the sequence $\alpha^{1/j}$ converges to 1. To see this, first note that the case $\alpha = 1$ is trivial, and the case $\alpha > 1$ implies the case $\alpha < 1$ (by taking reciprocals). So we concentrate on $\alpha > 1$. But then we have

$$1 < \alpha^{1/j} < j^{1/j}$$

when $j > \alpha$. Since $j^{1/j}$ tends to 1, Proposition 3.9 applies and the proof is complete. □

Example 3.18

Let $\lambda > 1$ and let α be real. Then the sequence

$$\left\{ \frac{j^\alpha}{\lambda^j} \right\}_{j=1}^\infty$$

converges to 0.

To see this, fix an integer $k > \alpha$ and consider $j > 2k$. (Notice that k is fixed once and for all but j will be allowed to tend to $+\infty$ at the appropriate moment.) Writing $\lambda = 1 + \mu, \mu > 0$, we have that

$$\lambda^j = (1+\mu)^j > \frac{j(j-1)(j-2)\cdots(j-k+1)}{k(k-1)(k-2)\cdots 2 \cdot 1} \mu^k \cdot 1^{j-k}.$$

Of course this comes from picking out the kth term of the binomial expansion for $(1 + \mu)^j$. Notice that since $j > 2k$ then each of the expressions $j, (j - 1), \ldots (j - k + 1)$ in the numerator on the right exceeds $j/2$. Thus

$$\lambda^j > \frac{j^k}{2^k \cdot k!} \cdot \mu^k$$

and

$$0 < \frac{j^\alpha}{\lambda^j} < j^\alpha \cdot \frac{2^k \cdot k!}{j^k \cdot \mu^k} = \frac{j^{\alpha-k} \cdot 2^k \cdot k!}{\mu^k}.$$

Since $\alpha - k < 0$, the right side tends to 0 as $j \to \infty$. □

Example 3.19

The sequence

$$\left\{ \left(1 + \frac{1}{j}\right)^j \right\}$$

converges. In fact it is monotone increasing and bounded above. Use the Binomial Expansion to prove this assertion. The limit of the sequence is the number that we shall later call e (in honor of Leonhard Euler, 1707–1783, who first studied it in detail). We shall study this sequence further in Proposition 4.20 of Section 4.4. □

Example 3.20

The sequence

$$\left\{ \left(1 - \frac{1}{j}\right)^j \right\}$$

converges to $1/e$, where the definition of e is given in the last example. More generally, the sequence

$$\left\{ \left(1 + \frac{x}{j} \right)^j \right\}$$

converges to e^x (here e^x is defined as in the discussion following Example 3.14 above). Exercise 14 asks you to prove these assertions. ▯

Exercises

3.1 Let $\{a_j\}, \{b_j\}$ be sequences. Prove that $\limsup(a_j + b_j) \leq \limsup a_j + \limsup b_j$. How are the lim inf's related? How is the quantity $(\limsup a_j) \cdot (\limsup b_j)$ related to $\limsup(a_j \cdot b_j)$? How are the lim inf's related?

3.2 Consider $\{a_j\}$ both as a sequence and as a set. How are the lim sup and the sup related? How are the lim inf and the inf related? Give examples.

3.3 Let $\{a_j\}$ be a sequence of positive numbers. How are the lim sup and lim inf of $\{a_j\}$ related to the lim sup and lim inf of $\{1/a_j\}$?

3.4 Prove parts (2) and (4) of Proposition 3.2.

3.5 Prove the following result, which is part of Corollary 3.8: Let S be a set of real numbers that is bounded above and let $\beta = \sup S$. For any $\epsilon > 0$ there is an element $s \in S$ such that $\beta - \epsilon < s \leq \beta$. (Remark: Notice that this lemma makes good intuitive sense: the elements of S should become arbitrarily close to the supremum β, otherwise there would be enough room to decrease the value of β and make the supremum even smaller.

3.6 Provide the details of the remark following the proof of the Bolzano–Weierstrass theorem.

3.7 Let $\{a_j\}$ be a sequence of real or complex numbers. Suppose that every subsequence has itself a subsequence that converges to a given number α. Prove that the full sequence converges to α.

3.8 Supply the details for Example 3.10.

3.9 Let $\emptyset$ be the empty set. Prove that $\sup \emptyset = -\infty$ and $\inf \emptyset = +\infty$.

3.10 Provide the details of the proof of Proposition 3.17 in case the lim sup is $+\infty$ or the lim inf is $-\infty$.

3.11 Provide the details of the assertion, made in the text, that the sequence $\{\cos j\}$ is dense in the interval $[-1, 1]$.

3.12 Prove Proposition 3.19.

3.13 Let α be a positive real number and let $p/q = m/n$ be two different representations of the same rational number r. Prove that

$$(\alpha^m)^{1/n} = (\alpha^p)^{1/q}.$$

Also prove that

$$(\alpha^{1/n})^m = (\alpha^m)^{1/n}.$$

If β is another positive real and γ is any real then prove that

$$(\alpha \cdot \beta)^\gamma = \alpha^\gamma \cdot \beta^\gamma.$$

3.14 Prove that the sequence

$$\left\{ \left(1 + \frac{x}{j} \right)^j \right\}.$$

converges to e^x for any real number x.

3.15 Discuss the convergence of the sequence $\{(1/j)^{1/j}\}$.

3.16 Find the lim sup and lim inf of the sequences

$$\{|\sin j|^{\sin j}\} \quad \text{and} \quad \{|\cos j|^{\cos j}\}.$$

3.17 Discuss the convergence of the sequence $\{(j^j)/(2j)!\}_{j=1}^{\infty}$.

3.18 How are the lim sup and lim inf of $\{a_j\}$ related to the lim sup and lim inf of $\{-a_j\}$?

3.19 Let $\{a_j\}$ be a real sequence. Prove that if

$$\liminf a_j = \limsup a_j$$

then the sequence $\{a_j\}$ converges. Prove the converse as well.

3.20 Let $a < b$ be real numbers. Give an example of a real sequence whose lim sup is b and whose lim inf is a.

3.21 Explain why we can make no sense of the concepts of lim sup and lim inf for complex sequences.

3.22 Let $\{a_j\}$ be a sequence of complex numbers. Suppose that for every pair of integers $N > M > 0$ it holds that $|a_M - a_{M+1}| + |a_{M+1} - a_{M+2}| + \cdots + |a_{N-1} - a_N| \leq 1$. Prove that $\{a_j\}$ converges.

3.23 Let $a_1, a_2 > 0$ and for $j \geq 3$ define $a_j = a_{j-1} + a_{j-2}$. Show that this sequence cannot converge to a finite limit.

3.24 Suppose a sequence $\{a_j\}$ has the property that for every natural number N there is a j_N such that $a_{j_N} = a_{j_N+1} = \cdots = a_{j_N+N}$. In other words, the sequence has arbitrarily long repetitive strings. Does it follow that the sequence converges?

3.25 Give an example of a single sequence of rational numbers with the property that for every real number α there is a subsequence converging to α.

3.26 Let $S = \{0, 1, 1/2, 1/3, 1/4, \ldots\}$. Give an example of a sequence $\{a_j\}$ with the property that for each $s \in S$ there is a subsequence converging to s, but no subsequence converges to any limit not in S.

3.27 Prove Proposition 3.9.

3.28 Give another proof of the Bolzano–Weierstrass theorem as follows. If $\{a_j\}$ is a bounded sequence let $b_j = \inf\{a_j, a_{j+1}, \ldots\}$. Then each b_j is finite, $b_1 \leq b_2 \leq \ldots$, and $\{b_j\}$ is bounded above. Now use Proposition 3.7.

4

Series of Numbers

4.1 Convergence of Series

In this section we will use standard summation notation:

$$\sum_{j=m}^{n} a_j \equiv a_m + a_{m+1} + \ldots + a_n.$$

A series is an infinite sum. The only way to handle an infinite process in mathematics is with a limit. This consideration leads to the following definition:

DEFINITION 4.1 *The formal expression*

$$\sum_{j=1}^{\infty} a_j,$$

*where the a_j's are real or complex numbers, is called a **series**. For $N = 1, 2, 3, \ldots$ the expression*

$$S_N = \sum_{j=1}^{N} a_j = a_1 + a_2 + \ldots a_N$$

*is called the Nth **partial sum** of the series. In case*

$$\lim_{N \to \infty} S_N$$

*exists and is finite we say that the series **converges**. Otherwise we say that the series **diverges**.*

Notice that the question of convergence of a series, which should be thought of as an *addition process*, reduces to a question about the *sequence* of partial sums.

Example 4.2
Consider the series

$$\sum_{j=1}^{\infty} 2^{-j}.$$

The Nth partial sum for this series is

$$S_N = 2^{-1} + 2^{-2} + \cdots + 2^{-N}.$$

In order to determine whether the sequence $\{S_N\}$ has a limit, we rewrite S_N as

$$S_N = \left(2^{-0} - 2^{-1}\right) + \left(2^{-1} - 2^{-2}\right) + \ldots$$
$$+ \left(2^{-N+1} - 2^{-N}\right).$$

The expression on the right of the last equation telescopes and we find that

$$S_N = 2^{-0} - 2^{-N}.$$

Thus

$$\lim_{N \to \infty} S_N = 2^{-0} = 1.$$

We conclude that the series converges.   □

Example 4.3
Let us examine the series

$$\sum_{j=1}^{\infty} \frac{1}{j}$$

for convergence or divergence. Now

$$S_1 = 1 = \frac{2}{2}$$
$$S_2 = 1 + \frac{1}{2} = \frac{3}{2}$$
$$S_4 = 1 + \frac{1}{2} + \left(\frac{1}{3} + \frac{1}{4}\right)$$
$$= 1 + \frac{1}{2} + \left(\frac{1}{4} + \frac{1}{4}\right) \geq 1 + \frac{1}{2} + \frac{1}{2} = \frac{4}{2}$$

$$S_8 = 1 + \frac{1}{2} + \left(\frac{1}{3} + \frac{1}{4}\right) + \left(\frac{1}{5} + \frac{1}{6} + \frac{1}{7} + \frac{1}{8}\right)$$

$$\geq 1 + \frac{1}{2} + \left(\frac{1}{4} + \frac{1}{4}\right) + \left(\frac{1}{8} + \frac{1}{8} + \frac{1}{8} + \frac{1}{8}\right)$$

$$= \frac{5}{2}.$$

In general this argument shows that

$$S_{2^k} \geq \frac{k+2}{2}.$$

The sequence of S_N's is increasing since the series contains only positive terms. The fact that the partial sums $S_1, S_2, S_4, S_8, \ldots$ increases without bound shows that the entire sequence of partial sums must increase without bound. We conclude that the series diverges.   □

Just as with sequences, we have a Cauchy criterion for series:

PROPOSITION 4.1
The series $\sum_{j=1}^{\infty} a_j$ converges if and only if for every $\epsilon > 0$ there is an integer $N > 1$ such that if $n \geq m > N$ then

$$\left| \sum_{j=m}^{n} a_j \right| < \epsilon. \tag{$*$}$$

*The condition $(*)$ is called the **Cauchy criterion for series**.*

PROOF   Suppose that the Cauchy criterion holds. Pick $\epsilon > 0$ and choose K so large that $(*)$ holds. If $n \geq m > K$ then

$$|S_n - S_m| = \left| \sum_{j=m+1}^{n} a_j \right| < \epsilon$$

by hypothesis. Thus the sequence $\{S_N\}$ is Cauchy in the sense discussed for sequences in Section 3.1. We conclude that the sequence $\{S_N\}$ converges; by definition, therefore, the series converges.

Conversely, if the series converges then, by definition, the sequence $\{S_N\}$ of partial sums converges. In particular the sequence $\{S_N\}$ must be Cauchy. Thus for any $\epsilon > 0$ there is a number $K > 0$ such that if $n \geq m > K$ then

$$|S_n - S_m| < \epsilon.$$

This just says that

$$\left| \sum_{j=m+1}^{n} a_j \right| < \epsilon,$$

and this last inequality is the Cauchy criterion for series. ∎

Example 4.4

Let us use the Cauchy criterion to verify that the series

$$\sum_{j=1}^{\infty} \frac{1}{j \cdot (j+1)}$$

converges.

Notice that if $n \geq m \geq 1$ then

$$\left| \sum_{j=m}^{n} \frac{1}{j \cdot (j+1)} \right| = \left(\frac{1}{m} - \frac{1}{m+1} \right) + \left(\frac{1}{m+1} - \frac{1}{m+2} \right) + \cdots$$
$$+ \left(\frac{1}{n} - \frac{1}{n+1} \right).$$

The sum on the right plainly telescopes and we have

$$\left| \sum_{j=m}^{n} \frac{1}{j \cdot (j+1)} \right| = \frac{1}{m} - \frac{1}{n+1}.$$

Let us choose N to be the next integer after $1/\epsilon$. Then for $n \geq m > N$ we may conclude that

$$\left| \sum_{j=m}^{n} \frac{1}{j \cdot (j+1)} \right| = \frac{1}{m} - \frac{1}{n+1} < \frac{1}{m} < \frac{1}{N} < \epsilon.$$

This is the desired conclusion. ▯

The next result gives a necessary condition for a series to converge. It is a useful device for detecting divergent series, although it can never tell us that a series converges.

PROPOSITION 4.2   THE ZERO TEST

If the series

$$\sum_{j=1}^{\infty} a_j$$

converges then the terms a_j tend to zero as $j \to \infty$.

PROOF   Since we are assuming that the series converges, then it must satisfy the Cauchy criterion. Let $\epsilon > 0$. Then there is an integer $N \geq 1$ such that if $n \geq m > N$ then

$$\left| \sum_{j=m}^{n} a_j \right| < \epsilon. \qquad (*)$$

We take $m > N$ and $n = m$. Then $(*)$ becomes

$$|a_m| < \epsilon.$$

But this is precisely the conclusion that we desire.   ∎

Example 4.5

The series $\sum_{j=1}^{\infty} (-1)^j$ must diverge, *even though its terms appear to be cancelling each other out.* The reason is that the summands do not tend to zero; hence the preceding proposition applies.

Write out several partial sums of this series to see more explicitly that the series diverges.   ☐

We conclude this section with a necessary and sufficient condition for convergence of a series of nonnegative terms. As with some of our other results on series, it amounts to little more than a restatement of a result on sequences.

PROPOSITION 4.3

A series

$$\sum_{j=1}^{\infty} a_j$$

with all $a_j \geq 0$ is convergent if and only if the sequence of partial sums is bounded.

PROOF   Notice that, because the summands are nonnegative, we have

$$S_1 = a_1 \leq a_1 + a_2 = S_2,$$

$$S_2 = a_1 + a_2 \leq a_1 + a_2 + a_3 = S_3,$$

and in general

$$S_N \leq S_N + a_{N+1} = S_{N+1}.$$

Thus the sequence $\{S_N\}$ of partial sums forms a monotone increasing sequence. We know that such a sequence is convergent to a finite limit if and only if it is bounded (see Section 3.1). This completes the proof.   ∎

Example 4.6
The series $\sum_{j=1}^{\infty} 1$ is divergent since the summands are nonnegative and the sequence of partial sums $\{S_N\} = \{N\}$ is unbounded.

Referring back to Example 4.3, we see that the series $\sum_{j=1}^{\infty} \frac{1}{j}$ diverges because its partial sums are unbounded.

We see from the first Example that the series $\sum_{j=1}^{\infty} 2^{-j}$ converges because its partial sums are all bounded above by 1.   ▯

It is frequently convenient to begin a series with summation at $j = 0$ or some other term instead of $j = 1$. All of our convergence results still apply to such a series because of the Cauchy criterion. In other words, the convergence or divergence of a series will depend only on the behavior of its "tail."

4.2   Elementary Convergence Tests

As previously noted, a series may converge because its terms diminish in size fairly rapidly (thus causing its partial sums to grow slowly) or it may converge because of cancellation among the terms. The tests that measure the first type of convergence are the most obvious and these are the "elementary" ones that we discuss in the present section.

PROPOSITION 4.4   THE COMPARISON TEST
Suppose that $\sum_{j=1}^{\infty} a_j$ is a convergent series of nonnegative terms. If $\{b_j\}$ are real or complex numbers and if $|b_j| \leq a_j$ for every j then the series $\sum_{j=1}^{\infty} b_j$ converges.

PROOF   Because the first series converges, its satisfies the Cauchy criterion for series. Hence, given $\epsilon > 0$, there is an N so large that if $n \geq m > N$ then

$$\left| \sum_{j=m}^{n} a_j \right| < \epsilon.$$

But then

$$\left| \sum_{j=m}^{n} b_j \right| \leq \sum_{j=m}^{n} |b_j| \leq \sum_{j=m}^{n} a_j < \epsilon.$$

It follows that the series $\sum b_j$ satisfies the Cauchy criterion for series. Therefore it converges.   ∎

COROLLARY 4.5

If $\sum_{j=1}^{\infty} a_j$ is as in the proposition and if $0 \leq b_j \leq a_j$ for every j then the series $\sum_{j=1}^{\infty} b_j$ converges.

PROOF   Obvious.   ∎

Example 4.7
The series $\sum_{j=1}^{\infty} 2^{-j} \sin j$ is seen to converge by comparing it with the series $\sum_{j=1}^{\infty} 2^{-j}$.   ▯

THEOREM 4.6   THE CAUCHY CONDENSATION TEST

Assume that $a_1 \geq a_2 \geq \ldots \geq a_j \geq \ldots 0$. The series

$$\sum_{j=1}^{\infty} a_j$$

converges if and only if the series

$$\sum_{k=1}^{\infty} 2^k \cdot a_{2^k}$$

converges.

PROOF First assume that the series $\sum_{j=1}^{\infty} a_j$ converges. Notice that, for each $k \geq 1$,

$$2^{k-1} \cdot a_{2^k} = a_{2^k} + a_{2^k} + \ldots + a_{2^k}$$

$$\leq a_{2^{k-1}+1} + a_{2^{k-1}+2} + \ldots a_{2^k}.$$

$$= \sum_{m=2^{k-1}+1}^{2^k} a_m.$$

Therefore

$$\sum_{k=1}^{N} 2^{k-1} \cdot a_{2^k} \leq \sum_{k=1}^{N} \sum_{m=2^{k-1}+1}^{2^k} a_m = \sum_{m=2}^{2^N} a_m.$$

Since the partial sums on the right are bounded (because the series of a_j's converges), so are the partial sums on the left. It follows that the series

$$\sum_{k=1}^{\infty} 2^k \cdot a_{2^k}$$

converges.

For the converse, assume that the series

$$\sum_{k=1}^{\infty} 2^k \cdot a_{2^k} \qquad (*)$$

converges. Observe that, for $k \geq 1$,

$$\sum_{m=2^{k-1}+1}^{2^k} a_m = a_{2^{k-1}+1} + a_{2^{k-1}+2} + \ldots + a_{2^k}$$

$$\leq a_{2^{k-1}} + a_{2^{k-1}} + \ldots + a_{2^{k-1}}$$

$$= 2^{k-1} \cdot a_{2^{k-1}}.$$

It follows that

$$\sum_{m=2}^{2^N} a_m = \sum_{k=1}^{N} \sum_{m=2^{k-1}+1}^{2^k} a_m$$

$$\leq \sum_{k=1}^{N} 2^{k-1} \cdot a_{2^{k-1}}.$$

By the hypothesis that the series $(*)$ converges, the partial sums on the right must be bounded. But then the partial sums on the left are bounded as well. By

Proposition 4.3, the series

$$\sum_j a_j$$

converges. ∎

Example 4.8

We apply the Cauchy condensation test to the harmonic series

$$\sum_{j=1}^{\infty} \frac{1}{j}.$$

It leads us to examine the series

$$\sum_{k=1}^{\infty} 2^k \cdot \frac{1}{2^k} = \sum_{k=1}^{\infty} 1.$$

Since the latter series diverges, the harmonic series diverges as well. ☐

PROPOSITION 4.7

Let α be a complex number. The series

$$\sum_{j=0}^{\infty} \alpha^j$$

*is called a **geometric series**. It converges if and only if $|\alpha| < 1$. In this circumstance, the sum of the series (that is, the limit of the partial sums) is $1/(1 - \alpha)$.*

PROOF   Let S_N denote the Nth partial sum of the geometric series. Then

$$\alpha \cdot S_N = \alpha(1 + \alpha + \alpha^2 + \ldots \alpha^N)$$
$$= \alpha + \alpha^2 + \ldots \alpha^{N+1}.$$

It follows that $\alpha \cdot S_N$ and S_N are nearly the same: in fact,

$$\alpha \cdot S_N + 1 - \alpha^{N+1} = S_N.$$

Solving this equation for the quantity S_N yields

$$S_N = \frac{1 - \alpha^{N+1}}{1 - \alpha}.$$

If $|\alpha| < 1$ then $\alpha^{N+1} \to 0$, hence the sequence of partial sums tends to the limit $1/(1 - \alpha)$. If $|\alpha| > 1$ then α^{N+1} diverges, hence the sequence of partial sums diverges. This completes the proof for $|\alpha| \neq 1$. But the divergence in case $|\alpha| = 1$ follows because the summands will not tend to zero. ∎

COROLLARY 4.8

The series

$$\sum_{j=1}^{\infty} \frac{1}{j^r}$$

converges if r is a real number that exceeds 1 and diverges otherwise.

PROOF We apply the Cauchy Condensation Test. This leads us to examine the series

$$\sum_{k=1}^{\infty} 2^k \cdot 2^{-kr} = \sum_{k=1}^{\infty} \left(2^{1-r}\right)^k.$$

This last is a geometric series, with the role of α played by the quantity 2^{1-r}. When $r > 1$ then $|\alpha| < 1$ so the series converges. Otherwise it diverges. ∎

THEOREM 4.9   THE ROOT TEST

Consider the series

$$\sum_{j=1}^{\infty} a_j$$

If

$$\limsup_{j \to \infty} |a_j|^{1/j} < 1$$

then the series converges.

PROOF Refer again to the discussion of the concept of limit superior in Chapter 3. By our hypothesis, there is a number $\beta < 1$ and an integer $N > 1$ such that for all $j > N$ it holds that

$$|a_j|^{1/j} < \beta.$$

In other words,

$$|a_j| < \beta^j.$$

Since $0 < \beta < 1$ the sum of the terms β^j on the right constitutes a convergent geometric series. By the Comparison Test, $\sum a_j$ converges. ∎

THEOREM 4.10   THE RATIO TEST

Consider a series

$$\sum_{j=1}^{\infty} a_j$$

of positive terms. If

$$\limsup_{j \to \infty} \left| \frac{a_{j+1}}{a_j} \right| < 1.$$

then the series converges.

PROOF   It is possible to supply a proof similar to that of the Root Test. We leave such a proof for the Exercises and instead supply an argument that relates the two tests in an interesting fashion.

Let

$$\lambda = \limsup_{j \to \infty} \left| \frac{a_{j+1}}{a_j} \right| < 1.$$

Select a real number μ such that $\lambda < \mu < 1$. By the definition of $\limsup$, there is an N so large that if $j \geq N$ then

$$\left| \frac{a_{j+1}}{a_j} \right| < \mu.$$

This may be rewritten as

$$|a_{j+1}| < \mu \cdot |a_j|, \qquad j \geq N.$$

Thus (much as in the proof of the Root Test) we have for $k \geq 1$ that

$$|a_{N+k}| \leq \mu \cdot |a_{N+k-1}| \leq \mu \cdot \mu \cdot |a_{N+k-2}| \leq \ldots \leq \mu^k \cdot |a_N|.$$

It is convenient to denote $N + k$ by $n, n \geq N$. Thus the last inequality reads

$$|a_n| < \mu^{n-N} \cdot |a_N|$$

or

$$|a_n|^{1/n} < \mu^{(n-N)/n} \cdot |a_N|^{1/n}.$$

Remembering that N has been fixed once and for all, we pass to the $\limsup$ as $n \to \infty$. The result is

$$\limsup_{n \to \infty} |a_n|^{1/n} \leq \mu.$$

Since $\mu < 1$, we find that our series satisfies the hypotheses of the Root Test. Hence it converges. ∎

REMARK 4.11 The proof of the Ratio Test shows that *if* a series passes the Ratio Test then it passes the Root Test (the converse is not true, as you will learn in the Exercises). Put another way, the Root Test is a better test than the Ratio Test because it will give information whenever the Ratio Test does and also (as we shall see) in some circumstances when the Ratio Test does not.

Why do we therefore learn the Ratio Test? The answer is that there are circumstances when the Ratio Test is easier to apply than the Root Test. ∎

Example 4.9

The series

$$\sum_{j=1}^{\infty} \frac{2^j}{j!}$$

is easily studied using the Ratio Test (recall that $j! \equiv j \cdot (j-1) \cdot \ldots 2 \cdot 1$). Indeed $a_j = 2^j / j!$ and

$$\left| \frac{a_{j+1}}{a_j} \right| = \frac{2^{j+1}/(j+1)!}{2^j/j!}.$$

We can perform the division to see that

$$\left| \frac{a_{j+1}}{a_j} \right| = \frac{2}{j+1}.$$

The lim sup of the last expression is 0. By the Ratio Test, the series converges.

Notice that in this example, while the Root Test applies in principle, it would be hard to use in practice. ▯

Example 4.10

We apply the Root Test to the series

$$\sum_{j=1}^{\infty} \frac{j^2}{2^j}.$$

Observe that

$$a_j = \frac{j^2}{2^j}$$

hence that

$$|a_j|^{1/j} = \frac{\left(j^{1/j}\right)^2}{2}.$$

As $j \to \infty$, we see that

$$\limsup_{j \to \infty} |a_j|^{1/j} = \frac{1}{2}.$$

By the Root Test, the series converges.   ⬚

It is natural to ask whether the Ratio and Root tests can detect divergence. Neither test is necessary and sufficient: there are series that elude the analysis of both tests. However, the arguments that we used to establish Theorems 4.9 and 4.10 can also be used to establish the following (the proofs are left as exercises):

THEOREM 4.12   THE ROOT TEST FOR DIVERGENCE
Consider the series

$$\sum_{j=1}^{\infty} a_j.$$

If

$$\limsup_{j \to \infty} |a_j|^{1/j} > 1$$

then the series diverges.

THEOREM 4.13   THE RATIO TEST FOR DIVERGENCE
Consider the series

$$\sum_{j=1}^{\infty} a_j.$$

If there is an $N > 0$ such that

$$\left| \frac{a_{j+1}}{a_j} \right| \geq 1, \qquad \forall j \geq N$$

then the series diverges.

In both the Root Test and the Ratio Test, if the lim sup is equal to one, then no conclusion is possible. The exercises give examples of series, some of which converge and some of which do not, in which these tests give lim sup equal to one.

4.3 Advanced Convergence Tests

In this section we consider some series convergence tests that depend on cancellation among the terms of the series. One of the most profound of these depends on a technique called *summation by parts*. You may wonder whether this process is at all related to the "integration by parts" procedure that you learned in calculus — it certainly has a similar form. Indeed it will turn out (and we shall see the details of this assertion as the book develops) that summing a series and performing an integration are two aspects of the same limiting process. The summation by parts method is merely our first glimpse of this relationship.

PROPOSITION 4.14   SUMMATION BY PARTS

Let $\{a_j\}_{j=0}^{\infty}$ and $\{b_j\}_{j=0}^{\infty}$ be two sequences of real or complex numbers. For $N = 0, 1, 2, \ldots$ set

$$A_N = \sum_{j=0}^{N} a_j$$

(we adopt the convention that $A_{-1} = 0$.) Then for any $0 \leq m \leq n < \infty$ it holds that

$$\sum_{j=m}^{n} a_j \cdot b_j = \left[A_n \cdot b_n - A_{m-1} \cdot b_m \right]$$

$$+ \sum_{j=m}^{n-1} A_j \cdot (b_j - b_{j+1}).$$

PROOF   We write

$$\sum_{j=m}^{n} a_j \cdot b_j = \sum_{j=m}^{n} \left(A_j - A_{j-1} \right) \cdot b_j$$

$$= \sum_{j=m}^{n} A_j \cdot b_j - \sum_{j=m}^{n} A_{j-1} \cdot b_j$$

$$= \sum_{j=m}^{n} A_j \cdot b_j - \sum_{j=m-1}^{n-1} A_j \cdot b_{j+1}$$

$$= \sum_{j=m}^{n-1} A_j \cdot (b_j - b_{j+1}) + A_n \cdot b_n - A_{m-1} \cdot b_m.$$

This is what we wish to prove.   ∎

Now we apply summation by parts to prove a convergence test due to Niels Henrik Abel (1802–1829).

THEOREM 4.15   ABEL'S CONVERGENCE TEST

Consider the series

$$\sum_{j=0}^{\infty} a_j \cdot b_j.$$

Suppose that

(**a**)   *the partial sums $A_N = \sum_{j=0}^{N} a_j$ form a bounded sequence;*

(**b**) $b_0 \geq b_1 \geq b_2 \geq \ldots;$

(**c**) $\lim_{j \to \infty} b_j = 0.$

Then the original series

$$\sum_{j=0}^{\infty} a_j \cdot b_j$$

converges.

PROOF   Suppose that the partial sums A_N are bounded in absolute value by a number K. Pick $\epsilon > 0$ and choose an integer N so large that $b_N < \epsilon/(2K)$. For $N \leq m \leq n < \infty$ we use the partial summation formula to write

$$\left| \sum_{j=m}^{n} a_j \cdot b_j \right| = \left| A_n \cdot b_n - A_{m-1} \cdot b_m + \sum_{j=m}^{n-1} A_j \cdot (b_j - b_{j+1}) \right|$$

$$\leq K \cdot |b_n| + K \cdot |b_m| + K \cdot \sum_{j=m}^{n-1} |b_j - b_{j+1}|.$$

Now we take advantage of the facts that $b_j \geq 0$ for all j and that $b_j \geq b_{j+1}$ for all j to estimate the last expression by

$$K \cdot \left[b_m + b_n + \sum_{j=m}^{n-1} \left(b_j - b_{j+1} \right) \right].$$

[Notice that the expressions $b_j - b_{j+1}, b_m$, and b_n are all nonnegative.] Now the sum collapses and the last line equals

$$K \cdot [b_m + b_n - b_n + b_m] = 2 \cdot K \cdot b_m.$$

By our choice of N the right side is smaller than ϵ. Thus our series satisfies the Cauchy criterion and therefore converges. ∎

Example 4.11 The Alternating Series Test

As a first application of Abel's convergence test, we examine alternating series. Consider a series of the form

$$\sum_{j=1}^{\infty} (-1)^j \cdot b_j, \qquad\qquad (*)$$

with $b_1 \geq b_2 \geq b_3 \geq \ldots \geq 0$ and $b_j \to 0$ as $j \to \infty$. We set $a_j = (-1)^j$ and apply Abel's test. We see immediately that all partial sums A_N are either -1 or 0. In particular, this sequence of partial sums is bounded. And the b_j's are monotone decreasing and tending to zero. By Abel's convergence test, the alternating series $(*)$ converges. ∎

PROPOSITION 4.16

Let $b_1 \geq b_2 \geq \ldots$ and assume that $b_j \to 0$. Consider the alternating series $\sum_{j=1}^{\infty} (-1)^j b_j$ as in the last example. It is convergent: let S be its sum. Then the partial sums S_N satisfy $|S - S_N| \leq b_{N+1}$.

PROOF   Observe that

$$|S - S_N| = |b_{N+1} - b_{N+2} + b_{N+3} - + \ldots|.$$

But

$$b_{N+2} - b_{N+3} + - \ldots \leq b_{N+2} + (-b_{N+3} + b_{N+3}) + (-b_{N+5} + b_{N+5}) + \ldots$$
$$= b_{N+2}$$

and

$$b_{N+2} - b_{N+3} + - \ldots \geq (b_{N+2} - b_{N+2}) + (b_{N+4} - b_{N+4}) + \ldots = 0.$$

It follows that

$$|S - S_N| \leq |b_{N+1}|,$$

as claimed. ∎

Example 4.12

Consider the series

$$\sum_{j=1}^{\infty}(-1)^j\frac{1}{j}.$$

Then the partial sum $S_{100} = -.688172$ is within 0.01 (in fact within $1/101$) of the full sum S, and the partial sum $S_{10000} = -.6930501$ is within 0.0001 (in fact within $1/10001$) of S.   ⬜

Example 4.13

Next we examine a series that is important in the study of Fourier analysis. Consider the series

$$\sum_{j=1}^{\infty}\frac{\sin j}{j}. \tag{$*$}$$

We already know that the series $\sum \frac{1}{j}$ diverges. However the expression $\sin j$ changes sign in a rather sporadic fashion. We might hope that the series $(*)$ converges because of cancellation of the summands. We take $a_j = \sin j$ and $b_j = 1/j$. Abel's test will apply if we can verify that the partial sums A_N of the a_j's are bounded. To see this we use a trick:

Observe that

$$\cos(j + 1/2) = \cos j \cdot \cos 1/2 - \sin j \cdot \sin 1/2$$

and

$$\cos(j - 1/2) = \cos j \cdot \cos 1/2 + \sin j \cdot \sin 1/2.$$

Subtracting these equations and solving for $\sin j$ yields that

$$\sin j = \frac{\cos(j - 1/2) - \cos(j + 1/2)}{2 \cdot \sin 1/2}.$$

 We conclude that

$$A_N = \sum_{j=1}^{N} a_j = \sum_{j=1}^{N}\frac{\cos(j - 1/2) - \cos(j + 1/2)}{2 \cdot \sin 1/2}.$$

Of course this sum collapses and we see that

$$A_N = \frac{-\cos(N + 1/2) + \cos 1/2}{2 \cdot \sin 1/2}.$$

Thus

$$|A_N| \leq \frac{2}{2 \cdot \sin 1/2} = \frac{1}{\sin 1/2},$$

independent of N.

Thus the hypotheses of Abel's test are verified and the series

$$\sum_{j=1}^{\infty} \frac{\sin j}{j}$$

converges. □

REMARK 4.17 It is interesting to notice that both the series

$$\sum_{j=1}^{\infty} \frac{|\sin j|}{j} \qquad \text{and} \qquad \sum_{j=1}^{\infty} \frac{\sin^2 j}{j}$$

diverge. The proofs of these assertions are left as exercises for you. ∎

We turn next to the topic of absolute and conditional convergence. A series of real or complex constants

$$\sum_{j=1}^{\infty} a_j$$

is said to be *absolutely convergent* if

$$\sum_{j=1}^{\infty} |a_j|$$

converges. We have:

PROPOSITION 4.18
If the series $\sum_{j=1}^{\infty} a_j$ is absolutely convergent then it is convergent.

PROOF This is an immediate corollary of the Comparison Test. ∎

DEFINITION 4.14 *A series $\sum_{j=1}^{\infty} a_j$ is said to be **conditionally convergent** if $\sum_{j=1}^{\infty} a_j$ converges, but it does not converge absolutely.*

We see that absolutely convergent series are convergent but the next example shows that the converse is not true.

Example 4.15
The series

$$\sum_{j=1}^{\infty} \frac{(-1)^j}{j} \tag{$*$}$$

converges by the Alternating Series Test. However it is not absolutely convergent because the harmonic series

$$\sum_{j=1}^{\infty} \frac{1}{j}$$

diverges. Thus the series $\sum (-1)^j/j$ converges conditionally.   ▯

There is a remarkable robustness result for absolutely convergent series that fails dramatically for conditionally convergent series. This result is enunciated in the next theorem. We first need a definition.

DEFINITION 4.16   *Let $\sum_{j=1}^{\infty} a_j$ be a given series. Let $\{p_j\}_{j=1}^{\infty}$ be a sequence in which every positive integer occurs once and only once (but not necessarily in the usual order). Then the series*

$$\sum_{j=1}^{\infty} a_{p_j}$$

*is said to be a **rearrangement** of the given series.*

THEOREM 4.19
If the series $\sum_{j=1}^{\infty} a_j$ of real numbers is absolutely convergent to a (limiting) sum ℓ then every rearrangement of the series converges also to ℓ. If the series $\sum_{j=1}^{\infty} b_j$ is conditionally convergent and if β is any real number or $\pm\infty$ then there is a rearrangement of the series such that its sequence of partial sums converges to β.

PROOF   We prove the first assertion here and explore the second in the Exercises.

Let us choose a rearrangement of the given series and denote it by $\sum_{j=1}^{\infty} a_{p_j}$. Pick $\epsilon > 0$. By the hypothesis that the original series converges absolutely we may choose an integer $N > 0$ such that $N < m \le n < \infty$ implies that

$$\sum_{j=m}^{n} |a_j| < \epsilon. \tag{$*$}$$

[The presence of the absolute values in the left side of this inequality will prove crucial in a moment.] Choose a positive integer M such that $M \ge N$ and the integers $1, \dots, N$ are all contained in the list $p_1, p_2, \dots, p_M$. If $K > M$ then the partial sum $\sum_{j=1}^{K} a_j$ will trivially contain the summands $a_1, a_2, \dots a_N$. Also the partial sum $\sum_{j=1}^{K} a_{p_j}$ will contain the summands $a_1, a_2, \dots a_N$. It follows that

$$\sum_{j=1}^{K} a_j - \sum_{j=1}^{K} a_{p_j}$$

will contain only summands *after* the Nth one in the original series. By inequality $(*)$ we may conclude that

$$\left| \sum_{j=1}^{K} a_j - \sum_{j=1}^{K} a_{p_j} \right| \leq \sum_{j=N+1}^{\infty} |a_j| \leq \epsilon.$$

We conclude that the rearranged series converges, and it converges to the same sum as the original series. ∎

4.4 Some Special Series

We begin with a series that defines a special constant of mathematical analysis.

DEFINITION 4.17   The series

$$\sum_{j=0}^{\infty} \frac{1}{j!},$$

where $j! \equiv j \cdot (j-1) \cdot (j-2) \cdots 1$ for $j \geq 1$ and $0! \equiv 1$, is convergent (by the Ratio Test, for instance). Its sum is denoted by the symbol e in honor of the Swiss mathematician Leonhard Euler who first studied it.

Like the number π, to be considered later in this book, the number e is one that arises repeatedly in a number of contexts in mathematics. It has many special properties. The first of these that we shall consider is that the definition that we have given for e is equivalent to another involving a sequence (this sequence was considered earlier in Example 3.19 of Section 3.4):

PROPOSITION 4.20
The limit

$$\lim_{n \to \infty} \left(1 + \frac{1}{n} \right)^n$$

exists and equals e.

PROOF   We need to compare the quantities

$$A_N \equiv \sum_{j=0}^{N} \frac{1}{j!} \quad \text{and} \quad B_N \equiv \left(1 + \frac{1}{N}\right)^N.$$

We use the binomial theorem to expand B_N:

$$B_N = 1 + \frac{N}{1} \cdot \frac{1}{N} + \frac{N \cdot (N-1)}{2 \cdot 1} \cdot \frac{1}{N^2} + \frac{N \cdot (N-1) \cdot (N-2)}{3 \cdot 2 \cdot 1} \cdot \frac{1}{N^3}$$

$$+ \ldots + \frac{N}{1} \cdot \frac{1}{N^{N-1}} + 1 \cdot \frac{1}{N^N}$$

$$= 1 + 1 + \frac{1}{2!} \cdot \frac{N-1}{N} + \frac{1}{3!} \cdot \frac{N-1}{N} \cdot \frac{N-2}{N} + \ldots$$

$$+ \frac{1}{(N-1)!} \cdot \frac{N-1}{N} \cdot \frac{N-2}{N} \cdots \frac{2}{N}$$

$$+ \frac{1}{N!} \cdot \frac{N-1}{N} \cdot \frac{N-2}{N} \cdots \frac{1}{N}$$

$$= 1 + 1 + \frac{1}{2!} \cdot \left(1 - \frac{1}{N}\right) + \frac{1}{3!} \cdot \left(1 - \frac{1}{N}\right) \cdot \left(1 - \frac{2}{N}\right) + \ldots$$

$$+ \frac{1}{(N-1)!} \cdot \left(1 - \frac{1}{N}\right) \cdot \left(1 - \frac{2}{N}\right) \cdots \left(1 - \frac{N-2}{N}\right)$$

$$+ \frac{1}{N!} \cdot \left(1 - \frac{1}{N}\right) \cdot \left(1 - \frac{2}{N}\right) \cdots \left(1 - \frac{N-1}{N}\right).$$

Notice that every expression that appears in this last equation is positive and does not exceed 1. Thus, for $0 \leq M \leq N$,

$$B_N \geq 1 + 1 + \frac{1}{2!} \cdot \left(1 - \frac{1}{N}\right) + \frac{1}{3!} \cdot \left(1 - \frac{1}{N}\right) \cdot \left(1 - \frac{2}{N}\right)$$

$$+ \ldots + \frac{1}{M!} \cdot \left(1 - \frac{1}{N}\right) \left(1 - \frac{2}{N}\right) \cdots \left(1 - \frac{M-1}{N}\right).$$

In this last inequality we hold M fixed and let N tend to infinity. The result is that

$$\liminf_{N \to \infty} B_N \geq 1 + 1 + \frac{1}{2!} + \frac{1}{3!} + \ldots + \frac{1}{M!} = A_M.$$

Now, as $M \to \infty$, the quantity A_M converges to e (by the *definition* of e). So we obtain

$$\liminf_{N \to \infty} B_N \geq e. \qquad (*)$$

On the other hand, our expansion for B_N allows us to observe that $B_N \leq A_N$. Thus

$$\limsup_{N \to \infty} B_N \leq e. \qquad (**)$$

Combining $(*)$ and $(**)$ we find that

$$e \leq \liminf_{N \to \infty} B_N \leq \limsup_{N \to \infty} B_N \leq e$$

hence that $\lim_{N \to \infty} B_N$ exists and equals e. This is the desired result. ∎

REMARK 4.21 The last proof illustrates the value of the concepts of $\liminf$ and $\limsup$. For we do not want to assume in advance that the limit of the expressions B_N exists, much less that the limit equals e. However the $\liminf$ and the $\limsup$ always exist. So we estimate those instead and find that they are equal and that they equal e. ∎

The next result tells us how rapidly the partial sums A_N of the series defining e converge to e. This is of theoretical interest, but will also be applied to determine the irrationality of e.

PROPOSITION 4.22
With A_N as above, we have that

$$0 < e - A_N < \frac{1}{N \cdot N!}.$$

PROOF Observe that

$$
\begin{aligned}
e - A_N &= \frac{1}{(N+1)!} + \frac{1}{(N+2)!} + \frac{1}{(N+3)!} + \cdots \\
&= \frac{1}{(N+1)!} \cdot \left(1 + \frac{1}{N+2} + \frac{1}{(N+2)(N+3)} + \cdots \right) \\
&< \frac{1}{(N+1)!} \cdot \left(1 + \frac{1}{N+1} + \frac{1}{(N+1)^2} + \cdots \right).
\end{aligned}
$$

Now the expression in parentheses is a geometric series. It sums to $(N+1)/N$. Since $A_N < e$, we have

$$e - A_N = |e - A_N|$$

hence

$$|e - A_N| < \frac{1}{N \cdot N!},$$

proving the result. ∎

Next we prove that e is an irrational number.

THEOREM 4.23

Euler's number e is irrational.

PROOF Suppose to the contrary that e is rational. Then $e = p/q$ for some positive integers p and q. By the preceding proposition,

$$0 < e - A_q < \frac{1}{q \cdot q!}$$

or

$$0 < q! \cdot (e - A_q) < \frac{1}{q}. \tag{$*$}$$

Now

$$e - A_q = \frac{p}{q} - \left(1 + 1 + \frac{1}{2!} + \frac{1}{3!} + \ldots + \frac{1}{q!} \right)$$

hence

$$q! \cdot (e - A_q)$$

is an integer. But then equation $(*)$ says that this integer lies between 0 and $1/q$. In particular, this integer lies strictly between 0 and 1. That, of course, is impossible. So e must be irrational. ∎

It is a general principle of number theory that a real number that can be approximated *too rapidly* by rational numbers (the degree of rapidity being measured in terms of powers of the denominators of the rational numbers) must be irrational. Under suitable conditions an even stronger conclusion holds: namely the number in question turns out to be *transcendental*. A transcendental number is one that is not the solution of any polynomial equation with integer coefficients.

The subject of transcendental numbers is explored in the Exercises. The Exercises also contain a sketch of a proof that e is transcendental.

Later in this book we shall discuss the number π and Euler's number γ. Both of these numbers arise in many contexts in mathematics. It is unknown whether γ is rational or irrational. The number π is known to be transcendental, but it is unknown whether $\pi + e$ is transcendental.

In recent years, questions about the the irrationality and transcendence of various numbers have become a matter of practical interest. For these properties prove to be useful in making and breaking secret codes, and in encrypting information so that it is accessible to some users but not to others.

Recall that, in Example 2.1, we proved that

$$S_N \equiv \sum_{j=1}^{N} j = \frac{N \cdot (N+1)}{2}.$$

We conclude this section with a method for summing higher powers of j.

Say that we wish to calculate

$$S_{k,N} \equiv \sum_{j=1}^{N} j^k$$

for some positive integer k exceeding 1. We may proceed as follows: write

$$
\begin{aligned}
(j+1)^{k+1} - j^{k+1} &= \left[j^{k+1} + (k+1) \cdot j^k + \frac{(k+1) \cdot k}{2} \cdot j^{k-1} \right.\\
&\quad \left. + \ldots + \frac{(k+1) \cdot k}{2} \cdot j^2 + (k+1) \cdot j + 1 \right]\\
&\quad - j^{k+1}\\
&= (k+1) \cdot j^k + \frac{(k+1) \cdot k}{2} \cdot j^{k-1} + \ldots\\
&\quad + \frac{(k+1) \cdot k}{2} \cdot j^2 + (k+1) \cdot j + 1.
\end{aligned}
$$

Summing from $j = 1$ to $j = N$ yields

$$
\begin{aligned}
\sum_{j=1}^{N} \left\{ (j+1)^{k+1} - j^{k+1} \right\} &= (k+1) \cdot S_{k,N} + \frac{(k+1) \cdot k}{2} \cdot S_{k-1,N} + \ldots\\
&\quad + \frac{(k+1) \cdot k}{2} \cdot S_{2,N} + (k+1) \cdot S_{1,N} + N.
\end{aligned}
$$

The sum on the left collapses to $(N+1)^{k+1} - 1$. We may solve for $S_{k,N}$ and obtain

$$
\begin{aligned}
S_{k,N} = \frac{1}{k+1} \cdot &\left[(N+1)^{k+1} - 1 - N - \frac{(k+1) \cdot k}{2} \cdot S_{k-1,N} \right.\\
&\quad \left. - \ldots - \frac{(k+1) \cdot k}{2} \cdot S_{2,N} - (k+1) \cdot S_{1,N} \right].
\end{aligned}
$$

We have succeeded in expressing $S_{k,N}$ in terms of $S_{1,N}, S_{2,N}, \ldots, S_{k-1,N}$.

Thus we may inductively obtain formulas for $S_{k,N}$, any k. It turns out that

$$S_{2,N} = \frac{N(N+1)(2N+1)}{6}$$

$$S_{3,N} = \frac{N^2(N+1)^2}{4}$$

$$S_{4,N} = \frac{(N+1)N(2N+1)(3N^2+3N-1)}{30}.$$

These formulas are treated in further detail in the Exercises.

4.5 Operations on Series

Some operations on series — such as addition, subtraction, and scalar multiplication — are straightforward. Others, such as multiplication, entail subtleties. This section treats these matters.

PROPOSITION 4.24

Let

$$\sum_{j=1}^{\infty} a_j \quad and \quad \sum_{j=1}^{\infty} b_j$$

be convergent series of real or complex numbers; assume that the series sum to limits α and β respectively. Then
   a) The series $\sum_{j=1}^{\infty}(a_j + b_j)$ converges to the
      limit $\alpha + \beta$.
   b) If c is a constant then the series $\sum_{j=1}^{\infty} c \cdot a_j$
      converges to $c \cdot \alpha$.

PROOF   We shall prove assertion (a) and leave the easier assertion (b) as an exercise.

Pick $\epsilon > 0$. Choose an integer N_1 so large that $n > N_1$ implies that the partial sum $S_n \equiv \sum_{j=1}^{n} a_j$ satisfies $|S_n - \alpha| < \epsilon/2$. Choose N_2 so large that $n > N_2$ implies that the partial sum $T_n \equiv \sum_{j=1}^{n} b_j$ satisfies $|T_n - \beta| < \epsilon/2$. If U_n is the nth partial sum of the series $\sum_{j=1}^{\infty}(a_j + b_j)$ and if $n > N_0 \equiv \max(N_1, N_2)$ then

$$|U_n - (\alpha + \beta)| \leq |S_n - \alpha| + |T_n - \beta| < \frac{\epsilon}{2} + \frac{\epsilon}{2} = \epsilon.$$

Thus the sequence $\{U_n\}$ converges to $\alpha + \beta$. This proves part (a). The proof of (b) is similar. ∎

In order to keep our discussion of multiplication of series as straightforward as possible, we deal at first with absolutely convergent series. It is convenient in this discussion to begin our sum at $j = 0$ instead of $j = 1$. If we wish to multiply

$$\sum_{j=0}^{\infty} a_j \quad \text{and} \quad \sum_{j=0}^{\infty} b_j,$$

then we need to specify what the partial sums of the product series should be. An obvious necessary condition that we wish to impose is that if the first series converges to α and the second converges to β then the product series, whatever we define it to be, should converge to $\alpha \cdot \beta$.

The naive method for defining the summands of the product series is to let $c_j = a_j \cdot b_j$. However, a glance at the product of two partial sums of the given series shows that such a definition would be ignoring the distributivity of addition.

Cauchy's idea was that the terms for the product series should be

$$c_n \equiv \sum_{j=0}^{n} a_j \cdot b_{n-j}.$$

This particular form for the summands can be easily motivated using power series considerations (which we shall provide later on). For now we concentrate on verifying that this "Cauchy product" of two series really works.

THEOREM 4.25

Let $\sum_{j=0}^{\infty} a_j$ and $\sum_{j=0}^{\infty} b_j$ be two absolutely convergent series that converge to limits α and β respectively. Define the series $\sum_{m=0}^{\infty} c_m$ with summands $c_m = \sum_{j=0}^{m} a_j \cdot b_{m-j}$. Then the series $\sum_{m=0}^{\infty} c_m$ converges to $\alpha \cdot \beta$.

PROOF Let $A_n, B_n,$ and C_n be the partial sums of the three series in question. We calculate that

$$C_n = (a_0 b_0) + (a_0 b_1 + a_1 b_0) + (a_0 b_2 + a_1 b_1 + a_2 b_0)$$

$$+ \cdots + \left(a_0 b_n + a_1 b_{n-1} + \cdots + a_{n-1} b_1 + a_n b_0 \right)$$

$$= a_0 \cdot B_n + a_1 \cdot B_{n-1} + a_2 \cdot B_{n-2} + \cdots + a_n \cdot B_0.$$

We set $\lambda_n = B_n - \beta$, each n, and rewrite the last line as

$$C_n = a_0(\beta + \lambda_n) + a_1(\beta + \lambda_{n-1}) + \cdots + a_n(\beta + \lambda_0)$$

$$= A_n \cdot \beta + \left[a_0 \lambda_n + a_1 \cdot \lambda_{n-1} + \cdots + a_n \cdot \lambda_0 \right].$$

Denote the expression in square brackets by the symbol ρ_n. Suppose that we could show that $\lim_{n\to\infty} \rho_n = 0$. Then we would have

$$
\begin{aligned}
\lim_{n\to\infty} C_n &= \lim_{n\to\infty} (A_n \cdot \beta + \rho_n) \\
&= \left(\lim_{n\to\infty} A_n \right) \cdot \beta + \left(\lim_{n\to\infty} \rho_n \right) \\
&= \alpha \cdot \beta + 0 \\
&= \alpha \cdot \beta.
\end{aligned}
$$

Thus it is enough to examine the limit of the expressions ρ_n.

Since $\sum_{j=1}^{\infty} a_j$ is absolutely convergent, we know that $A = \sum_{j=1}^{\infty} |a_j|$ is a finite number. Choose $\epsilon > 0$. Since $\sum_{j=1}^{\infty} b_j$ converges we know that $\lambda_n \to 0$. Thus we may choose an integer $N > 0$ such that $n > N$ implies that $|\lambda_n| < \epsilon$. Thus for $n = N + k, k > 0$, we may estimate

$$
\begin{aligned}
|\rho_{N+k}| &\leq |\lambda_0 a_{N+k} + \lambda_1 a_{N+k-1} + \cdots + \lambda_N a_k| \\
&\quad + |\lambda_{N+1} a_{k-1} + \lambda_{N+2} a_{k-2} + \cdots + \lambda_{N+k} a_0| \\
&\leq |\lambda_0 a_{N+k} + \lambda_1 a_{N+k-1} + + \cdots + \lambda_N a_k| \\
&\quad + \max_{p\geq 1}\{|\lambda_{N+p}|\} \cdot \big(|a_{k-1}| + |a_{k-2}| + \cdots + |a_0|\big) \\
&\leq (N+1) \cdot \max_{\ell \geq k}|a_\ell| \cdot \max_{0 \leq j \leq N}|\lambda_j| + \epsilon \cdot A.
\end{aligned}
$$

With N fixed, we let $k \to \infty$ in the last inequality. Since $\max_{\ell \geq k}|a_\ell| \to 0$, we find that

$$
\limsup_{n\to\infty} |\rho_n| \leq \epsilon \cdot A.
$$

Since $\epsilon > 0$ was arbitrary, we conclude that

$$
\lim_{n\to\infty} |\rho_n| \to 0.
$$

This completes the proof. ∎

Notice that, in the proof of the theorem, we really only used the fact that one of the given series was absolutely convergent, not that both were absolutely convergent. Some hypothesis of this nature is necessary, as the following example shows:

Example 4.18
Consider the Cauchy product of the two conditionally convergent series

$$
\sum_{j=0}^{\infty} \frac{(-1)^j}{\sqrt{j+1}} \quad \text{and} \quad \sum_{j=0}^{\infty} \frac{(-1)^j}{\sqrt{j+1}}.
$$

Observe that

$$c_m = \frac{(-1)^0 (-1)^m}{\sqrt{1}\sqrt{m+1}} + \frac{(-1)^1 (-1)^{m-1}}{\sqrt{2}\sqrt{m}} + \cdots$$

$$+ \frac{(-1)^m (-1)^0}{\sqrt{m+1}\sqrt{1}}$$

$$= \sum_{j=0}^{m} (-1)^m \frac{1}{\sqrt{(j+1)\cdot(m+1-j)}}.$$

However

$$(j+1)\cdot(m+1-j) \le (m+1)\cdot(m+1) = (m+1)^2.$$

Thus

$$|c_m| \ge \sum_{j=0}^{m} \frac{1}{m+1} = 1.$$

We thus see that the terms of the series $\sum_{m=0}^{\infty} c_m$ do not tend to zero, so the series cannot converge. $\square$

Exercises

4.1 Discuss convergence or divergence for each of the following series:

(a) $\sum_{j=1}^{\infty} \frac{(2^j)^2}{j!}$ (b) $\sum_{j=1}^{\infty} \frac{(2j)!}{(3j)!}$

(c) $\sum_{j=1}^{\infty} \frac{j!}{j^j}$ (d) $\sum_{j=1}^{\infty} \frac{(-1)^j}{3j^2 - 5j + 6}$

(e) $\sum_{j=1}^{\infty} \frac{2j-1}{3j^2 - 2}$ (f) $\sum_{j=1}^{\infty} \frac{2j-1}{3j^3 - 2}$

4.2 Let p be a polynomial with integer coefficients. Let $b_1 \ge b_2 \ge \ldots \ge 0$ and assume that $b_j \to 0$. Prove that if $(-1)^{p(j)}$ is not always positive and not always negative then in fact it will alternate in sign so that $\sum_{j=1}^{\infty} (-1)^{p(j)} \cdot b_j$ will converge.

4.3 If $b_j > 0$ for every j and if $\sum_{j=1}^{\infty} b_j$ converges then prove that $\sum_{j=1}^{\infty} (b_j)^2$ converges. Prove that the assertion is false if the positivity hypothesis is omitted. How about third powers?

4.4 If $b_j > 0$ for every j and if $\sum_{j=1}^{\infty} b_j$ converges then prove that $\sum_{j=1}^{\infty} \frac{1}{1+b_j}$ diverges.

4.5 If $b_j > 0$ for every j and if $\sum_{j=1}^{\infty} b_j$ converges then prove that $\sum_{j=1}^{\infty} \frac{b_j}{1+b_j}$ converges.

4.6 Let p be a polynomial with no constant term. If $b_j > 0$ for every j and if $\sum_{j=1}^{\infty} b_j$ converges then prove that the series $\sum_{j=1}^{\infty} p(b_j)$ converges.

4.7 Assume that $\sum_{j=1}^{\infty} b_j$ is an absolutely convergent series of real numbers. Let $s_j = \sum_{\ell=1}^{j} b_\ell$. Discuss convergence or divergence for the series $\sum_{j=1}^{\infty} s_j \cdot b_j$. Discuss convergence or divergence for the series $\sum_{j=1}^{\infty} \frac{b_j}{1+|s_j|}$.

4.8 If $b_j > 0$ for every j and if $\sum_{j=1}^{\infty} b_j$ diverges then define $s_j = \sum_{\ell=1}^{j} b_\ell$. Discuss convergence or divergence for the series $\sum_{j=1}^{\infty} \frac{b_j}{s_j}$.

4.9 Use induction to prove the formulas provided in the text for the sum of the first N perfect squares, the first N perfect cubes, and the first N perfect fourth powers.

4.10 Let $\sum_{j=1}^{\infty} b_j$ be a conditionally convergent series of real numbers. Let β be a real number. Prove that there is a rearrangement of the series that converges to β. (Hint: First observe that the positive terms of the given series must form a divergent series. Also, the negative terms form a divergent series. Now build the rearrangement by choosing finitely many positive terms whose sum "just exceeds" β. Then add on enough negative terms so that the sum is "just less than" β. Repeat this oscillatory procedure.)

4.11 Let $\sum_{j=1}^{\infty} a_j$ be a conditionally convergent series of complex numbers. Let S be the set of all possible complex numbers to which the various rearrangements could converge. What forms can S have? (Hint: experiment!)

4.12 Follow these steps to give another proof of the Alternating Series Test: a) Prove that the odd partial sums form an increasing sequence; b) prove that the even partial sums form a decreasing sequence; c) prove that every even partial sum majorizes all subsequent odd partial sums; d) use a pinching principle.

4.13 Examine the series

$$\left(\frac{1}{3} + \frac{1}{5}\right) + \left(\frac{1}{3^2} + \frac{1}{5^2}\right) + \left(\frac{1}{3^3} + \frac{1}{5^3}\right) + \frac{1}{3^4} + \frac{1}{5^4} + \cdots$$

Prove that the Root Test shows that the series converges while the Ratio Test gives no information.

4.14 Check that both the Root Test and the Ratio Test give no information for the series $\sum_{j=1}^{\infty} \frac{1}{j}$, $\sum_{j=1}^{\infty} \frac{1}{j^2}$. However one of these series is divergent and the other is convergent.

4.15 A real number s is called *algebraic* if it satisfies a polynomial equation of the form

$$a_0 + a_1 x + a_2 x^2 + \cdots + a_m x^m = 0$$

with the coefficients a_j being integers. Prove that if we replace the word "integers" in this definition with "rational numbers" then the set of algebraic numbers remains the same. Prove that $n^{p/q}$ is algebraic for any positive integers p, q, n.

4.16 Refer to Exercise 15 for terminology. A real number is called *transcendental* if it is not algebraic. Prove that the number of algebraic numbers is countable. Explain why this implies that the number of transcendental numbers is uncountable. Thus most real numbers are transcendental; however it is extremely difficult to verify that any particular real number is transcendental.

4.17 Refer to Exercises 15 and 16 for terminology. Provide the details of the following sketch of a proof that Euler's number e is transcendental. [Note: in this argument we use some simple ideas of calculus. These ideas will be treated in rigorous detail later in the book.] Seeking a contradiction, we suppose that the number e satisfies a polynomial equation of the form

$$a_0 + a_1 x + \ldots a_m x^m = 0$$

with integer coefficients a_j.

(a) We may assume that $a_0 \neq 0$.

(b) Let p be an odd prime that will be specified later. Define

$$g(x) = \frac{x^{p-1}(x-1)^p \cdots (x-m)^p}{(p-1)!}$$

and

$$G(x) = g(x) + g^{(1)}(x) + g^{(2)}(x) + \cdots g^{(mp+p-1)}(x).$$

(Here parenthetical exponents denote derivatives.) Verify that

$$|g(x)| < \frac{m^{mp+p-1}}{(p-1)!}.$$

(c) Check that

$$\frac{d}{dx}\left\{e^{-x}G(x)\right\} = -e^{-x}g(x)$$

and thus that

$$a_j \int_0^j e^{-x}g(x)\,dx = a_j G(0) - a_j e^{-j} G(j). \qquad (*)$$

(d) Multiply the last equation by e^j, sum from $j = 0$ to $j = m$, and use the polynomial equation that e satisfies to obtain that

$$\sum_{j=0}^{m} a_j e^j \int_0^j e^{-x}g(x)\,dx = -\sum_{j=0}^{m}\sum_{i=0}^{mp+p-1} a_j g^{(i)}(j). \qquad (**)$$

(e) Check that $g^{(i)}(j)$ is an integer for all values of i and all j from 0 to m inclusive.

(f) Referring to the last step, show that in fact $g^{(i)}(j)$ is an integer divisible by p *except* in the case that $j = 0$ and $i = p - 1$.

(g) Check that

$$g^{(p-1)}(0) = (-1)^p(-2)^p \cdots (-m)^p.$$

Conclude that $g^{(p-1)}(0)$ is not divisible by p if $p > m$.

(h) Check that if $p > |a_0|$ then the right side of equation $(**)$ consists of a sum of terms each of which is a multiple of p *except* for the term $-a_0 g^{(p-1)}(0)$. It follows that the sum on the right side of $(**)$ is a nonzero integer.

(i) Use equation $(*)$ to check that, provided p is chosen sufficiently large, the left side of $(**)$ satisfies

$$\left|\sum_{j=0}^{m} a_j e^j \int_0^j e^{-x}g(x)\,dx\right| \leq \left\{\sum_{j=0}^{m}|a_j|\right\} e^m \frac{(m^{m+2})^{p-1}}{(p-1)!} < 1.$$

(j) The last two steps contradict each other.

This proof is from [NIV].

4.18 Prove Theorem 4.12.

4.19 Prove Theorem 4.13.

4.20 Let $\sum_{j=1}^{\infty} a_j$ and $\sum_{j=1}^{\infty} b_j$ be convergent series of positive real numbers. Discuss division of these two series. Use the idea of the Cauchy product.

4.21 Let $\sum_{j=1}^{\infty} a_j$ and $\sum_{j=1}^{\infty} b_j$ be convergent series of positive real numbers. Discuss convergence of $\sum_{j=1}^{\infty} a_j b_j$.

4.22 What can you say about the convergence or divergence of

$$\sum_{j=1}^{\infty} \frac{(2j+3)^{1/2} - (2j)^{1/2}}{j^{1/2}} \ ?$$

4.23 If $b_j > 0$ and $\sum_{j=1}^{\infty} b_j$ converges then prove that

$$\sum_{j=1}^{\infty} (b_j)^{1/2} \cdot \frac{1}{j^{\alpha}}$$

converges for any $\alpha > 1/2$. Give an example to show that the assertion is false if $\alpha = 1/2$.

4.24 Let a_j be a sequence of real numbers. Define

$$m_j = \frac{a_1 + a_2 + \ldots a_j}{j}.$$

Prove that if $\lim_{j \to \infty} a_j = \ell$ then $\lim_{j \to \infty} m_j = \ell$. Give an example to show that the converse is not true.

4.25 Imitate the proof of the Root Test to give a direct proof of the Ratio Test.

4.26 Prove that

$$\sum_{j=1}^{\infty} \frac{|\sin j|}{j} \qquad \text{and} \qquad \sum_{j=1}^{\infty} \frac{\sin^2 j}{j}$$

are both divergent series.

4.27 Prove Proposition 4.24(b).

4.28 Let $\sum_{j=1}^{\infty} a_j$ be a divergent series of positive terms. Prove that there exist numbers $b_j, 0 < b_j < a_j$, such that $\sum_{j=1}^{\infty} b_j$ diverges.

Similarly, let $\sum_{j=1}^{\infty} c_j$ be a convergent series of positive terms. Prove that there exist numbers $d_j, 0 < c_j < d_j$, such that $\sum_{j=1}^{\infty} d_j$ converges.

Thus we see that there is no "smallest" divergent series and no "largest" convergent series.

4.29 Let $\sum_j a_j$ and $\sum_j b_j$ be series. Prove that if there is a constant $C > 0$ such that

$$\frac{1}{C} \le \left| \frac{a_j}{b_j} \right| \le C$$

for all j large then either both series diverge or both series converge.

4.30 Refer to Exercise 15 for terminology. Prove that the sum of two algebraic numbers is algebraic. Prove that the product of two algebraic numbers is algebraic.

4.31 Is the analogue of Exercise 30 true for transcendental numbers (see Exercise 16)?

5

Basic Topology

5.1 Open and Closed Sets

To specify a topology on a set is to describe certain subsets that will play the role of neighborhoods. These sets are called *open sets*.

In what follows, we will use "interval notation": If $a \leq b$ are real numbers then we define

$$(a, b) = \{x \in \mathbb{R} : a < x < b\},$$

$$[a, b] = \{x \in \mathbb{R} : a \leq x \leq b\},$$

$$[a, b) = \{x \in \mathbb{R} : a \leq x < b\},$$

$$(a, b] = \{x \in \mathbb{R} : a < x \leq b\}.$$

Intervals of the form (a, b) are called *open*. Those of the form $[a, b]$ are called *closed*. The other two are termed *half-open* or *half-closed*.

DEFINITION 5.1 *A set $U \subseteq \mathbb{R}$ is called **open** if for each $x \in U$ there is an $\epsilon > 0$ such that the interval $(x - \epsilon, x + \epsilon)$ is contained in U.*

Example 5.2
The set $U = \{x \in \mathbb{R} : |x - 3| < 2\}$ is open. To see this, choose a point $x \in U$. Let $\epsilon = 2 - |x - 3| > 0$. Then we claim that the interval $I = (x - \epsilon, x + \epsilon) \subseteq U$.

For if $t \in I$ then

$$|t - 3| \leq |t - x| + |x - 3|$$

$$< \epsilon + |x - 3|$$

$$= (2 - |x - 3|) + |x - 3| = 2.$$

But this means that $t \in U$.

We have shown that $t \in I$ implies $t \in U$. Therefore $I \subseteq U$. It follows from the definition that U is open. □

The way to think about the definition of open set is that a set is open when none of its elements is at the "edge" of the set — each element is surrounded by other elements of the set, indeed a whole interval of them. The remainder of this section will make these comments precise.

PROPOSITION 5.1

If U_α are open sets, for α in some (possibly uncountable) index set A, then

$$U = \bigcup_{\alpha \in A} U_\alpha$$

is open.

PROOF Let $x \in U$. By definition of union, the point x must lie in some U_α. But U_α is open. Therefore there is an interval $I = (x - \epsilon, x + \epsilon)$ such that $I \subseteq U_\alpha$. Therefore certainly $I \subseteq U$. This proves that U is open. ∎

PROPOSITION 5.2

If $U_1, U_2, \ldots, U_k$ are open sets then the set

$$V = \bigcap_{j=1}^{k} U_j$$

is also open.

PROOF Let $x \in V$. Then $x \in U_j$ for each j. Since each U_j is open there is for each j a positive number ϵ_j such that $I_j = (x - \epsilon_j, x + \epsilon_j)$ lies in U_j. Set $\epsilon = \min\{\epsilon_1, \ldots, \epsilon_k\}$. Then $\epsilon > 0$ and $(x - \epsilon, x + \epsilon) \subseteq U_j$ for every j. But that just means that $(x - \epsilon, x + \epsilon) \subseteq V$. Therefore V is open. ∎

Notice the difference between these two propositions: arbitrary unions of open sets are open. But in order to guarantee that an intersection of open sets is still open we had to assume that we were only intersecting finitely many such sets. To understand this matter bear in mind the example of the open sets

$$U_j = \left(-\frac{1}{j}, \frac{1}{j}\right) \quad , \quad j = 1, 2, \ldots .$$

The intersection of the sets U_j is the singleton $\{0\}$, which is not open.

The same analysis as in the first example shows that if $a < b$ then the interval (a, b) is an open set. On the other hand, intervals of the form $(a, b]$ or $[a, b)$ or $[a, b]$ are *not* open. In the first instance, the point b is the center of no interval $(b - \epsilon, b + \epsilon)$ contained in $(a, b]$. Think about the other two intervals to understand why they are not open.

We are now in a position to give a complete description of all open sets.

PROPOSITION 5.3

Let $U \subseteq \mathbb{R}$ be an open set. Then there are countably many pairwise disjoint open intervals I_j such that

$$U = \bigcup_{j=1}^{\infty} I_j.$$

PROOF Assume that U is an open subset of the real line. We define an equivalence relation on the set U. The resulting equivalence classes will be the open intervals I_j.

Let a and b be elements of U. We say that a is related to b if all real numbers between a and b are also elements of U. It is obvious that this relation is both reflexive and symmetric. For transitivity notice that if a is related to b and b is related to c then (assuming that a, b, c are distinct) one of the numbers a, b, c must lie between the other two. Assume for simplicity that $a < b < c$. Then all numbers between a and c lie in U, for all such numbers are either between a and b or between b and c or are b itself. (The other possible orderings of a, b, c are left for you to consider.)

Thus we have an equivalence relation on the set U. Call the equivalence classes $\{U_\alpha\}_{\alpha \in A}$. We claim that each U_α is an open interval. In fact if a, b are elements of some U_α then all points between a and b are in U. But then a moment's thought shows that each of those "in between" points is related to both a and b. Therefore all points between a and b are elements of U_α. We conclude that U_α is an interval. Is it an *open* interval?

Let $x \in U_\alpha$. Then $x \in U$ so that there is an open interval $I = (x - \epsilon, x + \epsilon)$ contained in U. But x is related to all the elements of I; it follows that $I \subseteq U_\alpha$. Therefore U_α is open.

We have exhibited the set U as a union of open intervals. These intervals are pairwise disjoint because they arise as equivalence classes of an equivalence relation. Finally, each of these open intervals contains a (different) rational number (why?). Therefore there can be at most countably many of the intervals U_α. ∎

DEFINITION 5.3 *A subset $F \subseteq \mathbb{R}$ is called **closed** if the complement $\mathbb{R} \setminus F$ is open.*

Example 5.4

An interval of the form $[a, b] = \{x : a \leq x \leq b\}$ is closed. For its complement is $(-\infty, a) \cup (b, \infty)$, which is the union of two open intervals.

The finite set $A = \{-4, -2, 5, 13\}$ is closed because its complement is

$$(-\infty, -4) \cup (-4, -2) \cup (-2, 5) \cup (5, 13) \cup (13, \infty)$$

which is open.

The set $B = \{1, 1/2, 1/3, 1/4, \ldots\} \cup \{0\}$ is closed, for its complement is the set

$$(-\infty, 0) \cup \left\{ \bigcup_{j=1}^{\infty} (1/(j+1), 1/j) \right\} \cup (1, \infty),$$

which is open.

Verify for yourself that if the point 0 is omitted from the set B then the set is no longer closed. ▯

PROPOSITION 5.4

If E_α are closed sets, for α in some (possibly uncountable) index set A, then

$$E = \bigcap_{\alpha \in A} E\alpha$$

is closed.

PROOF This is just the contrapositive of Proposition 5.1 above: if U_α is the complement of E_α, each α, then U_α is open. Then $U = \cup U_\alpha$ is also open. But then

$$E = \cap E_\alpha = \cap^c (U_\alpha) = {}^c (\cup U_\alpha) = {}^c U$$

is closed. ∎

The fact that the set B in the last example is closed, but that $B \setminus \{0\}$ is not, is placed in perspective by the next proposition:

PROPOSITION 5.5

Let S be a set of real numbers. Then S is closed if and only if every Cauchy sequence $\{s_j\}$ of elements of S has a limit that is also an element of S.

PROOF First suppose that S is closed and let $\{s_j\}$ be a Cauchy sequence in S. We know, since the reals are complete, that there is an element $s \in \mathbb{R}$ such that $s_j \to s$. The point of this half of the proof is to see that $s \in S$. If this statement were false then $s \in T = \mathbb{R} \setminus S$. But T must be open since it is the complement of a closed set. Thus there is an $\epsilon > 0$ such that the interval $I = (s - \epsilon, s + \epsilon) \subseteq T$. This means that no element of S lies in I. In particular,

$|s - s_j| \geq \epsilon$ for every j. This contradicts the statement that $s_j \to s$. We conclude that $s \in S$.

Conversely, assume that every Cauchy sequence in S has its limit in S. If S were not closed then its complement would not be open. Hence there would be a point $t \in \mathbb{R} \setminus S$ with the property that no interval $(t - \epsilon, t + \epsilon)$ lies in $\mathbb{R} \setminus S$. In other words, $(t - \epsilon, t + \epsilon) \cap S \neq \emptyset$ for every $\epsilon > 0$. Thus for $j = 1, 2, 3, \ldots$ we may choose a point $s_j \in (t - 1/j, t + 1/j) \cap S$. It follows that $\{s_j\}$ is a sequence of elements of S that converge to $t \in \mathbb{R} \setminus S$. That contradicts our hypothesis. We conclude that S must be closed. ∎

Let S be a subset of $\mathbb{R}$. A point x is called an *accumulation point* of S if every neighborhood of x contains infinitely many distinct elements of S. In particular, x is an accumulation point of S if it is the limit of a nonconstant sequence in S. The last proposition tells us that closed sets are characterized by the property that they contain all of their accumulation points.

5.2 Further Properties of Open and Closed Sets

Let $S \subseteq \mathbb{R}$ be a set. We call $b \in \mathbb{R}$ a *boundary point* of S if every nonempty neighborhood $(b - \epsilon, b + \epsilon)$ contains both points of S and points of $\mathbb{R} \setminus S$. We denote the set of boundary points of S by ∂S.

A boundary point b might lie in S and might lie in the complement of S. The next example serves to illustrate the concept.

Example 5.5

Let S be the interval $(0, 1)$. Then no point of $(0, 1)$ is in the boundary of S since every point of $(0, 1)$ has a neighborhood that lies inside $(0, 1)$. Also, no point of the complement of $[0, 1]$ lies in the boundary of S for a similar reason. Indeed, the only candidates for elements of the boundary of S are 0 and 1. The point 0 *is* an element of the boundary since every neighborhood $(0 - \epsilon, 0 + \epsilon)$ contains the points $(0, \epsilon) \subseteq S$ and points $(-\epsilon, 0] \subseteq \mathbb{R} \setminus S$. A similar calculation shows that 1 lies in the boundary of S.

Now consider the set $T = [0, 1]$. Certainly there are no boundary points in $(0, 1)$, for the same reason as in the first paragraph. And there are no boundary points in $\mathbb{R} \setminus [0, 1]$, since that set is open. Thus the only candidates for elements of the boundary are 0 and 1. As in the first paragraph, these are both indeed boundary points for T.

Notice that neither of the boundary points of S lie in S while both of the boundary points of T lie in T. ☐

Example 5.6

The boundary of the set $\mathbb{Q}$ is the entire real line. For if x is any element of $\mathbb{R}$ then every interval $(x - \epsilon, x + \epsilon)$ contains both rational numbers and irrational numbers. ▯

DEFINITION 5.7 *Let $S \subseteq \mathbb{R}$. A point $s \in S$ is called an **interior point** of S if there is an $\epsilon > 0$ such that the interval $(s - \epsilon, s + \epsilon)$ lies in S.*

*A point $t \in S$ is called an **isolated point** of S if there is an $\epsilon > 0$ such that the intersection of the interval $(t - \epsilon, t + \epsilon)$ with S is just the singleton $\{t\}$.*

By the definitions given here, an isolated point of a set $S \subseteq \mathbb{R}$ is a boundary point. For any interval $(s - \epsilon, s + \epsilon)$ contains a point of S (namely s itself) and points of $\mathbb{R} \setminus S$ (since s is isolated).

PROPOSITION 5.6

Let $S \subseteq \mathbb{R}$. Then each point of S is either an interior point or a boundary point.

PROOF Fix $s \in S$. If s is not an interior point then no open interval centered at s contains only elements of s. Thus any interval centered at s contains an element of S (namely s itself) and also contains points of $\mathbb{R} \setminus S$. Thus s is a boundary point of S. ∎

Example 5.8

Let $S = [0, 1]$. Then the interior points of S are the elements of $(0, 1)$. The boundary points of S are the points 0 and 1. The set S has no isolated points. Let $T = \{1, 1/2, 1/3, \ldots\} \cup \{0\}$. Then the points $1, 1/2, 1/3, \ldots$ are isolated points of T. Every element of T is a boundary point, and there are no others. ▯

REMARK 5.7 Observe that the interior points of a set S are *elements* of S — by their very definition. Also, isolated points of S are elements of S. However, a boundary point of S may or may not be an element of S.

If x is an accumulation point of S then every open neighborhood of x contains infinitely many elements of S. Hence x is either a boundary point of S or an interior point of S; it *cannot* be an isolated point of S. ∎

PROPOSITION 5.8

Let S be a subset of the real numbers. Then the boundary of S equals the boundary of $\mathbb{R} \setminus S$.

PROOF Obvious. ∎

The next theorem allows us to use the concept of boundary to distinguish open sets from closed sets.

THEOREM 5.9

A closed set contains all of its boundary points. An open set contains none of its boundary points.

PROOF Let S be closed and let x be an element of its boundary. If every neighborhood of x contains points of S *other than x itself* then x is an accumulation point, hence $x \in S$. If not every neighborhood of x contains points of S other than x itself, then there is an $\epsilon > 0$ such that $[(x, x - \epsilon) \cup (x, x + \epsilon)] \cap S = \emptyset$. The only way that x can be an element of ∂S in this circumstance is if $x \in S$. That is what we wished to prove.

For the other half of the theorem notice that if T is open then ^cT is closed. But then ^cT will contain all its boundary points, which are the same as the boundary points of T itself. Thus T can contain none of its boundary points. ∎

PROPOSITION 5.10

Every non-isolated boundary point of a set S is an accumulation point of the set S.

PROOF This proof is treated in the Exercises. ∎

The converse of the last proposition is false. For example, *every* point of the set $[0, 1]$ is an accumulation point of the set, yet only 0 and 1 are boundary points.

DEFINITION 5.9 *A subset S of the real numbers is called* **bounded** *if there is a positive number M such that $|s| \leq M$ for every element s of S.*

The next result is one of the great theorems of nineteenth century analysis. It is essentially a restatement of the Bolzano–Weierstrass Theorem of Section 3.2.

THEOREM 5.11

Every bounded, infinite subset of $\mathbb{R}$ has an accumulation point.

PROOF Let S be a bounded, infinite set of real numbers. Let $\{a_j\}$ be a sequence of distinct elements of S. By Theorem 3.11, there is a subsequence $\{a_{j_k}\}$ that converges to a limit α. Then α is an accumulation point of S. ∎

COROLLARY 5.12

*Let $S \subseteq \mathbb{R}$ be a closed and bounded set. If $\{a_j\}$ is any sequence in S then there is a Cauchy subsequence $\{a_{j_k}\}$ that converges **to an element of** S.*

PROOF Merely combine the Bolzano–Weierstrass theorem with Proposition 5.5 of the last section. ∎

5.3 Compact Sets

Compact sets are sets (usually infinite) that share many of the most important properties of finite sets. They play an important role in real analysis.

DEFINITION 5.10 A set $S \subseteq \mathbb{R}$ is called **compact** if every sequence in S has *a subsequence that converges **to an element of** S.*

PROPOSITION 5.13

A set is compact if and only if it is closed and bounded.

PROOF That a closed, bounded set has the property of compactness is the content of Corollary 5.12 of the last section.

Now let S be a set that is compact. If S is not bounded, then there is an element s_1 of S that has absolute value larger than 1. Also there must be an element s_2 of S that has absolute value larger than 2. Continuing, we find elements $s_j \in S$ satisfying

$$|s_j| > j$$

for each j. But then the sequence $\{s_j\}$ cannot be Cauchy. This contradiction shows that S must be bounded.

If S is compact but S is not closed, then there is a point x that is the limit of a sequence $\{s_j\} \subseteq S$ but that is not itself in S. But every sequence in S is, by definition of "compact," supposed to have a subsequence converging *to an element of* S. For the sequence $\{s_j\}$ that we are considering, x is the only candidate for the limit of a subsequence. Thus it must be that $x \in S$. That contradiction establishes that S is closed. ∎

In the abstract theory of topology (where there is no notion of distance), sequences cannot be used to characterize topological properties. Therefore a different definition of compactness is used. For interest's sake, and for future use, we now show that the definition of compactness that we have been discussing is equivalent to the one used in topology theory. First we need a definition.

DEFINITION 5.11   *Let S be a subset of the real numbers. A collection of open sets $\{\mathcal{O}_\alpha\}_{\alpha \in A}$ (each $\mathcal{O}_\alpha$ is an open set of real numbers) is called an **open covering** of S if*

$$\bigcup_{\alpha \in A} \mathcal{O}_\alpha \supseteq S.$$

Example 5.12
The collection $\mathcal{C} = \{(1/j, 1)\}_{j=1}^{\infty}$ is an open covering of the interval $I = (0, 1)$. Observe, however, that no subcollection of $\mathcal{C}$ covers I.

The collection $\mathcal{D} = \{(1/j, 1)\}_{j=1}^{\infty} \cup \{(-1/5, 1/5)\} \cup \{(4/5, 6/5)\}$ is an open covering of the interval $J = [0, 1]$. However, not all the elements $\mathcal{D}$ are actually needed to cover J. In fact

$$(-1/5, 1/5) \,, \ (1/6, 1) \,, \ (4/5, 6/5)$$

cover the interval J.   ☐

It is the distinction displayed in this example that distinguishes compact sets from the point of view of topology. We need another definition:

DEFINITION 5.13   *If $\mathcal{C}$ is an open covering of a set S and if $\mathcal{D}$ is another open covering of S such that each element of $\mathcal{D}$ is also an element of $\mathcal{C}$ then we call $\mathcal{D}$ a **subcovering** of $\mathcal{C}$.*

*We call $\mathcal{D}$ a **finite subcovering** if $\mathcal{D}$ has just finitely many elements.*

Example 5.14
The collection of intervals

$$\mathcal{C} = \{(j - 1, j + 1)\}_{j=1}^{\infty}$$

is an open covering of the set $S = [5, 9]$. The collection

$$\mathcal{D} = \{(j - 1, j + 1)\}_{j=5}^{\infty}$$

is a subcovering.

However the collection

$$\mathcal{E} = \{(4,6),(5,7),(6,8),(7,9),(8,10)\}$$

is a *finite* subcovering. ☐

THEOREM 5.14   THE HEINE–BOREL THEOREM

A set $S \subseteq \mathbb{R}$ is compact if and only if every open covering $\mathcal{C} = \{\mathcal{O}_\alpha\}_{\alpha \in A}$ of S has a finite subcovering.

PROOF   Assume that S is a compact set and let $\mathcal{C} = \{\mathcal{O}_\alpha\}_{\alpha \in A}$ be an open covering of S.

By Proposition 5.13, S is closed and bounded. Therefore it holds that $a = \inf S$ is a finite real number and an element of S. Likewise, $b = \sup S$ is a finite real number and an element of S. Write $I = [a,b]$. Set

$$\mathcal{A} = \{x \in I : \mathcal{C} \text{ contains a finite subcover that covers } S \cap [a,x]\}.$$

Then $\mathcal{A}$ is nonempty since $a \in \mathcal{A}$. Let $t = \sup \mathcal{A}$. Then some element $\mathcal{O}_0$ of $\mathcal{C}$ contains t. Let s be an element of $\mathcal{O}_0$ to the left of t. Then, by the definition of t, s is an element of $\mathcal{A}$. So there is a finite subcovering $\mathcal{C}'$ of $\mathcal{C}$ that covers $[a,s] \cap S$. But then $\mathcal{D} = \mathcal{C}' \cup \{\mathcal{O}_0\}$ covers $[a,t] \cap S$, showing that $t = \sup \mathcal{A}$ lies in $\mathcal{A}$. But in fact $\mathcal{D}$ even covers points to the right of t. Thus t cannot be the supremum of $\mathcal{A}$ unless $t = b$.

We have learned that t must be the point b itself and that therefore $b \in \mathcal{A}$. But that says that $S \cap [a,b] = S$ can be covered by finitely many of the elements of $\mathcal{C}$. That is what we wished to prove.

For the converse, assume that every open covering of S has a finite subcovering. Let $\{a_j\}$ be a sequence in S. Assume, seeking a contradiction, that the sequence has no subsequence that converges to an element of S. This must mean that for every $s \in S$ there is an $\epsilon_s > 0$ such that no element of the sequence satisfies $0 < |a_j - s| < \epsilon_s$. Let $I_s = (s - \epsilon_s, s + \epsilon_s)$. The collection $\mathcal{C} = \{I_s\}$ is then an open covering of the set S. By hypothesis, there exists a finite subcovering $I_{s_1}, \ldots I_{s_k}$ of open intervals that cover S. But each I_{s_ℓ} could only contain at most one element of the sequence $\{a_j\}$ — namely it is possible that $s_\ell = a_j$ for some j. We conclude that the sequence has only finitely many distinct elements, a clear contradiction. Thus the sequence does have a convergent subsequence. ∎

Example 5.15

If $A \subseteq B$ and both sets are nonempty then $A \cap B = A \neq \emptyset$. A similar assertion holds when intersecting *finitely many* nonempty sets $A_1 \supseteq A_2 \supseteq \ldots \supseteq A_k$; in this circumstance we have $\cap_{j=1}^{k} A_j = A_k$.

However it is possible to have infinitely many nonempty nested sets with null intersection. An example is the sets $I_j = (0, 1/j)$. Certainly $I_j \supseteq I_{j+1}$ for all j yet

$$\bigcap_{j=1}^{\infty} I_j = \emptyset.$$

By contrast, if we take $K_j = [0, 1/j]$ then

$$\bigcap_{j=1}^{\infty} K_j = \{0\}.$$

The next proposition shows that compact sets have the intuitively appealing property of the K_j's rather than the unsettling property of the I_j's. ⬜

PROPOSITION 5.15

Let

$$K_1 \supseteq K_2 \supseteq \ldots \supseteq K_j \ldots$$

be nonempty compact sets of real numbers. Set

$$\mathcal{K} = \bigcap_{j=1}^{\infty} K_j.$$

Then $\mathcal{K}$ is compact and $\mathcal{K} \neq \emptyset$.

PROOF Each K_j is closed and bounded hence $\mathcal{K}$ is closed and bounded. Therefore $\mathcal{K}$ is compact. Let $x_j \in K_j$, each j. Then $\{x_j\} \subseteq K_1$. By compactness, there is a convergent subsequence $\{x_{j_k}\}$ with limit $x_0 \in K_1$. However, $\{x_{j_k}\}_{k=2}^{\infty} \subseteq K_2$. Thus $x_0 \in K_2$. Similar reasoning shows that $x_0 \in K_m$ for all $m = 1, 2, \ldots$. In conclusion, $x_0 \in \cap_j K_j = \mathcal{K}$. ∎

5.4 The Cantor Set

In this section we describe the construction of a remarkable subset of $\mathbb{R}$ with many pathological properties. It only begins to suggest the richness of the structure of the real number system.

We begin with the unit interval $S_0 = [0, 1]$. We extract from S_0 its open middle third; thus $S_1 = S_0 \setminus (1/3, 2/3)$. Observe that S_1 consists of two closed intervals of equal length $1/3$.

Now we construct S_2 from S_1 by extracting from each of its two intervals the middle third: $S_2 = [0, 1/9] \cup [2/9, 3/9] \cup [6/9, 7/9] \cup [8/9, 1]$. The figure shows S_2.

$$0 \qquad\qquad\qquad\qquad\qquad\qquad\qquad\qquad\qquad 1$$

Continuing in this fashion, we construct S_{j+1} from S_j by extracting the middle third from each of its component subintervals. We define the Cantor set C to be

$$C = \bigcap_{j=1}^{\infty} S_j.$$

Notice that each of the sets S_j is closed and bounded, hence compact. By Proposition 5.15 of the last section, C is therefore not empty. The set C is closed and bounded, hence compact.

PROPOSITION 5.16
The Cantor set C has zero length, in the sense that $[0, 1] \setminus C$ has length 1.

PROOF In the construction of S_1, we removed from the unit interval one interval of length 3^{-1}. In constructing S_2, we further removed two intervals of length 3^{-2}. In constructing S_j, we removed 2^{j-1} intervals of length 3^{-j}. Thus the total length of the intervals removed from the unit interval is

$$\sum_{j=1}^{\infty} 2^{j-1} \cdot 3^{-j}.$$

This last equals

$$\frac{1}{3} \sum_{j=0}^{\infty} \left(\frac{2}{3}\right)^j.$$

The geometric series sums easily and we find that the total length of the intervals removed is

$$\frac{1}{3} \left(\frac{1}{1 - 2/3}\right) = 1.$$

Thus the Cantor set has length zero because its complement in the unit interval has length one. ∎

PROPOSITION 5.17

The Cantor set is uncountable.

PROOF   We assign to each element of the Cantor set a "label" consisting of a sequence of 1's and 2's that identifies its location in the set.

Fix an element x in the Cantor set. Then certainly x is in S_1. If x is in the left half of S_1, then the first digit in the "label" of x is 1; otherwise it is 2. Likewise $x \in S_2$. By the first part of this argument, it is either in the left half S_{21} of S_2 (when the first digit in the label is 1) or the right half S_{22} of S_2 (when the first digit of the label is 2). Whichever of these is correct, that half will consist of two intervals of length 3^{-2}. If x is in the leftmost of these two intervals then the second digit of the "label" of x is 1. Otherwise the second digit is 2. Continuing in this fashion, we may assign to x an infinite sequence of 1's and 2's.

Conversely, if $a, b, c, \ldots$ is a sequence of 1's and 2's, then we may locate a unique corresponding element y of the Cantor set. If the first digit is a one then y is in the left half of S_1; otherwise y is in the right half of S_1. Likewise the second digit locates y within S_2, and so forth.

Thus we have a one-to-one correspondence between the Cantor set and the collection of all infinite sequences of ones and twos. [Notice that we are in effect thinking of the point assigned to a sequence $c_1 c_2 c_3 \ldots$ of 1's and 2's as the limit of the points assigned to $c_1, c_1 c_2, c_1 c_2 c_3, \ldots$. Thus we are using the fact that C is closed.] However, as we learned in Chapter 1, the set of all infinite sequences of ones and twos is uncountable. Thus the Cantor set is uncountable.

∎

The Cantor set is quite thin (it has zero length) but it is large in the sense that it has uncountably many elements. Also it is compact. The next result reveals a surprising, and not generally well known, property of this "thin" set:

THEOREM 5.18

Let C be the Cantor set and define

$$S = \{x + y : x \in C, y \in C\}.$$

Then $S = [0, 2]$.

PROOF   We sketch the proof here and treat the details in the exercises.

Since $C \subseteq [0, 1]$ it is clear that $S \subseteq [0, 2]$. For the reverse inclusion, fix an element $t \in [0, 2]$. Our job is to find two elements c and d in C such that $c + d = t$.

First observe that $\{x + y : x \in S_1, y \in S_1\} = [0, 2]$. Therefore there exist $x_1 \in S_1$ and $y_1 \in S_1$ such that $x_1 + y_1 = t$.

Similarly, $\{x + y : x \in S_2, y \in S_2\} = [0, 2]$. Therefore there exist $x_2 \in S_2$ and $y_2 \in S_2$ such that $x_2 + y_2 = t$.

Continuing in this fashion we may find for each j numbers x_j and y_j such that $x_j, y_j \in S_j$ and $x_j + y_j = t$. Of course $\{x_j\} \subseteq C$ and $\{y_j\} \subseteq C$ hence there are subsequences $\{x_{j_k}\}$ and $\{y_{j_k}\}$ that converge to real numbers c and d. Since C is compact, we can be sure that $c \in C$ and $d \in C$. But the operation of addition respects limits, thus we may pass to the limit as $k \to \infty$ in the equation

$$x_{j_k} + y_{j_k} = t$$

to obtain

$$c + d = t.$$

Therefore $[0, 2] \subseteq \{x + y : x \in C\}$. This completes the proof. ∎

In the exercises at the end of the chapter we shall explore constructions of other Cantor sets, some of which have zero length and some of which have positive length. We shall also consider other ways to construct the Cantor set that we have been discussing in this section.

Observe that, whereas any open set is the union of open intervals, the existence of the Cantor set shows us that there is no such structure theorem for closed sets. In fact, closed intervals are atypically simple when considered as examples of closed sets.

5.5   Connected and Disconnected Sets

Let S be a set of real numbers. We say that S is *disconnected* if it is possible to find a pair of open sets U and V such that

$$U \cap S \neq \emptyset, V \cap S \neq \emptyset,$$

$$(U \cap S) \cap (V \cap S) = \emptyset,$$

and

$$S = (U \cap S) \cup (V \cap S).$$

If no such U and V exist then we call S *connected*.

Example 5.16
The set $T = \{x \in \mathbb{R} : |x| < 1, x \neq 0\}$ is disconnected. For take $U = \{x : x < 0\}$ and $V = \{x : x > 0\}$. Then

$$U \cap T = \{x : -1 < x < 0\} \neq \emptyset$$

and

$$V \cap T = \{x : 0 < x < 1\} \neq \emptyset.$$

Also $(U \cap T) \cap (V \cap T) = \emptyset$. Clearly $T = (U \cap T) \cup (V \cap T)$, hence T is disconnected. ☐

Example 5.17

The set $X = [-1, 1]$ is connected. To see this, suppose to the contrary that there exist open sets U and V such that $U \cap X \neq \emptyset$, $V \cap X \neq \emptyset$, $(U \cap X) \cap (V \cap X) = \emptyset$, and

$$X = (U \cap X) \cup (V \cap X).$$

Choose $a \in U \cap X$ and $b \in V \cap X$. Set

$$\alpha = \sup \left(U \cap [a, b] \right).$$

Then $\alpha \leq b$. Now $[a, b] \subseteq X$ hence $U \cap [a, b]$ is disjoint from V. But ^cV is closed, hence $\alpha \notin V$. It follows that $\alpha < b$.

If $\alpha \in U$ then, because U is open, there exists an $\tilde{\alpha} \in U$ such that $\alpha < \tilde{\alpha} < b$. This would mean that we chose α incorrectly. Hence $\alpha \notin U$. But $\alpha \notin U$ and $\alpha \notin V$ means $\alpha \notin X$. On the other hand, α is the supremum of a subset of X (since $a \in X, b \in X$, and X is an interval). Since X is a closed interval, we conclude that $\alpha \in X$. This contradiction shows that X must be connected. ☐

With small modifications, the discussion in the last example demonstrates that any closed interval is connected (Exercise 11). Also (see Exercise 12), we may similarly see that any open interval or half-open interval is connected. In fact the converse is true as well:

THEOREM 5.19

If S is a connected subset of $\mathbb{R}$ then S is an interval.

PROOF If S is not an interval then there exist $a \in S$, $b \in S$ and a point t between a and b such that $t \notin S$. Define $U = \{x \in \mathbb{R} : x < t\}$ and $V = \{x \in \mathbb{R} : t < x\}$. Then U and V are open, nonempty, and disjoint, $U \cap S \neq \emptyset$, $V \cap S \neq \emptyset$, and

$$S = (U \cap S) \cup (V \cap S).$$

Thus S is disconnected.

We have proved the contrapositive of the statement of the theorem, hence we are finished. ∎

The Cantor set is not connected; indeed it is disconnected in a special sense. Call a set S *totally disconnected* if for each distinct $x \in S$, $y \in S$, there exist disjoint open sets U and V such that $x \in U$, $y \in V$, and $S = (U \cap S) \cup (V \cap S)$.

PROPOSITION 5.20

The Cantor set is totally disconnected.

PROOF   Let $x, y \in C$ be distinct and assume that $x < y$. Set $\delta = |x - y|$. Choose j so large that $3^{-j} < \delta$. Then $x, y \in S_j$, but x and y cannot both be in the same interval of S_j (since the intervals will be of length equal to 3^{-j}). It follows that there is a point t between x and y that is not an element of S_j, hence certainly not an element of C. Set $U = \{s : s < t\}$ and $V = \{s : s > t\}$. Then $x \in U \cap C$, hence $U \cap C \neq \emptyset$; likewise $V \cap C \neq \emptyset$. Also $(U \cap C) \cap (V \cap C) = \emptyset$. Finally, $C = (C \cap U) \cup (C \cap V)$. Thus C is totally disconnected.   ∎

5.6   Perfect Sets

A set $S \subseteq \mathbb{R}$ is called *perfect* if it is closed and if every point of S is an accumulation point of S. The property of being perfect is a rather special one: it means that the set has no isolated points.

Obviously a closed interval $[a, b]$ is perfect. After all, a point x in the interior of the interval is surrounded by an entire open interval $(x - \epsilon, x + \epsilon)$ of elements of the interval; moreover, a is the limit of elements from the right and b is the limit of elements from the left.

Perhaps more surprising is that the Cantor set, *a totally disconnected set*, is perfect. It is certainly closed. Now fix $x \in C$. Then certainly $x \in S_1$. Thus x is in one of the two intervals comprising S_1. One (or perhaps both) of the endpoints of that interval does not equal x. Call that endpoint a_1. Likewise $x \in S_2$. Therefore x lies in one of the intervals of S_2. Choose an endpoint a_2 of that interval which does not equal x. Continuing in this fashion, we construct a sequence $\{a_j\}$. Notice that *each of the elements of this sequence lies in the Cantor set* (why?). Finally, $|x - a_j| \leq 3^{-j}$ for each j. Therefore x is the limit of the sequence. We have thus proved that the Cantor set is perfect.

The fundamental theorem about perfect sets tells us that such a set must be rather large. We have

THEOREM 5.21

A nonempty perfect set must be uncountable.

PROOF   Let S be a perfect set. Since S has accumulation points, it cannot be finite. Therefore it is either countable or uncountable.

Seeking a contradiction, we suppose that S is countable. Write $S = \{s_1, s_2, \ldots\}$. Set $U_1 = (s_1 - 1, s_1 + 1)$. Then U_1 is a neighborhood of s_1. Now s_1 is a limit point of S so there must be infinitely many elements of S lying in U_1. We select a bounded open interval U_2 such that $\bar{U}_2 \subseteq U_1$, $\bar{U}_2$ does not contain s_1, and U_2 *does* contain some element of S.

Continuing in this fashion, assume that $s_1, \ldots, s_j$ have been selected and choose a bounded interval U_{j+1} such that (i) $\bar{U}_{j+1} \subseteq U_j$, (ii) $s_j \notin \bar{U}_{j+1}$, and (iii) U_{j+1} contains some element of S.

Observe that each set $V_j = \bar{U}_j \cap S$ is closed and bounded, hence compact. Also each V_j is nonempty by construction but V_j does not contain s_{j-1}. It follows that $V = \cap_j V_j$ cannot contain s_1 (since V_2 does not), cannot contain s_2 (since V_3 does not), indeed cannot contain any element of S. Hence V, being a subset of S, is empty. But V is the decreasing intersection of nonempty compact sets, hence cannot be empty!

This contradiction shows that S cannot be countable. So it must be uncountable. ∎

COROLLARY 5.22

If $a < b$ then the closed interval $[a, b]$ is uncountable.

PROOF The interval $[a, b]$ is perfect. ∎

We also have a new way of seeing that the Cantor set is uncountable, since it is perfect:

COROLLARY 5.23

The Cantor set is uncountable.

Exercises

5.1 Let S be any set of real numbers. Denote by $\overset{\circ}{S}$ its set of interior points. Call this set the *interior* of S. Prove that $\overset{\circ}{S}$ is open. Prove that S is open if and only if S equals its interior.

5.2 Let S be any set of real numbers. Define the *closure* of S, denoted $\bar{S}$, to be the union of S and its boundary. Prove that $S \subseteq \bar{S}$. Prove that $\bar{S}$ is a closed set. Prove that $\bar{S} \setminus \overset{\circ}{S}$ is the boundary of S.

5.3 Let K be a compact set and let U be an open set that contains K. Prove that there is an $\epsilon > 0$ such that if $k \in K$ then the interval $(k - \epsilon, k + \epsilon)$ is contained in U.

5.4 Let S be any set and $\epsilon > 0$. Define $T = \{t \in \mathbb{R} : |t - s| < \epsilon \text{ for some } s \in S\}$. Prove that T is open.

falls thru hole of the = to

5.5 Let S be any set and define $V = \{t \in \mathbb{R} : |t - s| \leq 1 \text{ for some } s \in S\}$. Is V necessarily closed?

5.6 Fix the sequence $a_j = 3^{-j}, j = 1, 2, \ldots$. Consider the set S of all sums

$$\sum_{j=1}^{\infty} \mu_j a_j,$$

where each μ_j is one of the numbers 0 or 2. Show that S is the Cantor set. If s is an element of S, $s = \sum \mu_j a_j$, and if $\mu_j = 0$ for all j sufficiently large, then show that s is an endpoint of one of the intervals in one of the sets S_j that were used to construct the Cantor set in the text.

5.7 Discuss which sequences a_j of positive numbers could be used as in Exercise 6 to construct sets that are like the Cantor set.

5.8 Let us examine the proof that $\{x + y : x \in C, y \in C\}$ equals $[0, 2]$ more carefully.

    (a) Prove for each j that $\{x + y : x \in S_j, y \in S_j\}$ equals the interval $[0, 2]$.

    (b) Explain how the subsequences $\{x_{j_k}\}$ and y_{j_k} can be chosen to satisfy $x_{j_k} + y_{j_k} = t$. Observe that it is important for the proof that the index j_k be the same for both subsequences.

    (c) Formulate a suitable statement concerning the assertion that the binary operation of addition "respects limits" as required in the argument in the text. Prove this statement and explain how it allows us to pass to the limit in the equation $x_{j_k} + y_{j_k} = t$.

5.9 Use the characterization of the Cantor set from Exercise 6 to give a new proof of the fact that $\{x + y : x \in C, y \in C\}$ equals the interval $[0, 2]$.

5.10 See Exercises 1 and 2 for terminology. Call a set S *robust* if it is the closure of its interior. Which sets of reals are robust?

5.11 Imitate the example in the text to prove that any closed interval is connected.

5.12 Imitate the example in the text to prove that any open interval or half-open interval is connected.

5.13 Construct a Cantor-like set by removing the middle *fifth* from the unit interval, removing the middle fifth of each of the remaining intervals, and so on. What is the length of the set that you construct in this fashion? Is it uncountable? Is it perfect? Is it different from the Cantor set constructed in the text?

5.14 Refer to Exercise 13. Construct a Cantor set by removing the middle third from the unit interval, removing the middle ninth (*not* the middle third as in the text) from each of the remaining intervals, removing the middle twenty-seventh from each of the remaining intervals after that, and so on. The Cantor-like set that results should have positive length. What is that length? Does this Cantor set have the other properties of the Cantor set constructed in the text?

5.15 Refer to Exercises 13 and 14. Let $0 < \alpha < 1$. Construct a Cantor-like set that has length α. Verify that this set has all the properties of the Cantor set that was discussed in the text.

5.16 Let $X_1, X_2, \ldots$ each be perfect sets and suppose that $X_1 \supseteq X_2 \supseteq \ldots$. Set $X = \cap_j X_j$. Is X perfect?

5.17 Give an example of nonempty *closed* sets $X_1 \supseteq X_2 \supseteq \ldots$ such that $\cap_j X_j = \emptyset$.

5.18 Give an example of nonempty closed sets $X_1 \subseteq X_2 \ldots$ such that $\cup_j X_j$ is open.

5.19 Give an example of open sets $U_1 \supseteq U_2 \ldots$ such that $\cap_j U_j$ is closed and nonempty.

5.20 Give an example of a totally disconnected set $S \subseteq [0, 1]$ such that $\bar{S} = [0, 1]$.

5.21 What is the interior of the Cantor set? What is the boundary of the Cantor set?

5.22 Write the real line as the union of two totally disconnected sets.

5.23 Construct a sequence α of real numbers with the property that for every $x \in \mathbb{R}$ there is a subsequence of α that converges to x.

5.24 Let $S_1, S_2, \ldots$ be closed sets and assume that $\cup_j S_j = \mathbb{R}$. Prove that at least one of the sets S_j has nonempty interior. (Hint: Use an idea from the proof that perfect sets are uncountable.)

5.25 Let K be a compact set and let $\{U_\alpha\}_{\alpha \in A}$ be an open covering of K. Prove that there is an $\epsilon > 0$ such that if $k \in K$ then the interval $(k - \epsilon, k + \epsilon)$ lies in some U_α.

5.26 Let $U_1 \subseteq U_2 \ldots$ be open sets and assume that each of these sets has bounded, nonempty complement. Prove that $\cup_j U_j \neq \mathbb{R}$.

5.27 Exhibit a countable collection of open sets U_j such that each open set $\mathcal{O} \subseteq \mathbb{R}$ can be written as a union of some of the sets U_j.

5.28 Let S be a nonempty set of real numbers. A point x is called a *condensation point* of S if every neighborhood of x contains uncountably many points of S. Prove that the set of condensation points of S is closed. Is it necessarily nonempty? Is it nonempty when S is uncountable?

   If T is an uncountable set then show that the set of its condensation points is perfect.

5.29 Prove that any closed set can be written as the union of a perfect set and a countable set. (Hint: Refer to Exercise 28.)

5.30 Let S be an uncountable subset of $\mathbb{R}$. Prove that S must have infinitely many accumulation points. Must it have uncountably many?

5.31 Let S be a compact set and T a closed set of real numbers. Assume that $S \cap T = \emptyset$. Prove that there is a number $\delta > 0$ such that $|s - t| > \delta$ for every $s \in S$ and every $t \in T$. Prove that the assertion is false if we only assume that S is closed.

5.32 Prove that the assertion of Exercise 31 is false if we assume that S and T are both open.

5.33 Let S be any set and define, for $x \in \mathbb{R}$,

$$\text{dis}(x, S) = \inf\{|x - s| : s \in S\}.$$

Prove that if $x \notin \bar{S}$ then $\text{dis}(x, S) > 0$. If $x, y \in \mathbb{R}$ then prove that

$$|\text{dis}(x, S) - \text{dis}(y, S)| \leq |x - y|.$$

5.34 Let S be a set of real numbers. If S is not open then must it be closed? If S is not closed then must it be open?

5.35 Prove Proposition 5.10.

6

Limits and Continuity of Functions

6.1 Definition and Basic Properties of the Limit of a Function

In this chapter we are going to treat some topics that you have seen before in your calculus class. However we shall use the deep properties of the real numbers that we have developed in this text to obtain important new insights. Therefore you should *not* think of this chapter as review. Look at the concepts introduced here with the power of your new understanding of analysis.

DEFINITION 6.1 *Let $E \subseteq \mathbb{R}$ be a set and let f be a real-valued function with domain E. Fix a point $P \in \mathbb{R}$ that is either in E or is an accumulation point of E. We say that*

$$\lim_{E \ni x \to P} f(x) = \ell,$$

with ℓ a real number, if for each $\epsilon > 0$ there is a $\delta > 0$ such that when $x \in E$ and $0 < |x - P| < \delta$ then

$$|f(x) - \ell| < \epsilon.$$

The definition makes precise the notion that we can force $f(x)$ to be just as close as we please to ℓ by making x sufficiently close to P. Notice that the definition puts the condition $0 < |x - P| < \delta$ on x — in other words we do not look at $x = P$, but rather at x *near* to P.

Also observe that we only consider the limit of f at a point P that is not in the interior of the complement of E. In the Exercises you will be asked to discuss why it would be nonsensical to use the above definition to study limits at such points.

Example 6.2

Let $E = \mathbb{R} \setminus \{0\}$ and

$$f(x) = x \cdot \sin(1/x) \quad \text{if} \quad x \in E.$$

Then $\lim_{x \to 0} f(x) = 0$. To see this, let $\epsilon > 0$. Choose $\delta = \epsilon$. If $0 < |x - 0| < \delta$ then

$$|f(x) - 0| = |x \cdot \sin(1/x)| \leq |x| < \delta = \epsilon,$$

as desired. Thus the limit exists and equals 0. ☐

Example 6.3

Let $E = \mathbb{R}$ and

$$g(x) = \begin{cases} 1 & \text{if} \quad x \text{ is rational} \\ 0 & \text{if} \quad x \text{ is irrational.} \end{cases}$$

Then $\lim_{x \to P} g(x)$ does not exist for any point P of E.

To see this, fix $P \in \mathbb{R}$. Seeking a contradiction, assume that there is a limiting value ℓ for g at P. If this is so then we take $\epsilon = 1/2$ and we can find a $\delta > 0$ such that $0 < |x - P| < \delta$ implies

$$|g(x) - \ell| < \epsilon = \frac{1}{2}. \tag{*}$$

If we take x to be rational then $(*)$ says that

$$|1 - \ell| < \frac{1}{2}, \tag{**}$$

while if we take x irrational then $(*)$ says that

$$|0 - \ell| < \frac{1}{2}. \tag{***}$$

But then the triangle inequality gives that

$$|1 - 0| = |(1 - \ell) + (\ell - 0)|$$
$$\leq |1 - \ell| + |\ell - 0|,$$

which by $(**)$ and $(***)$ is

$$< 1.$$

This contradiction, that $1 < 1$, allows us to conclude that the limit does not exist at P. ☐

PROPOSITION 6.1

Let f be a function with domain E, and let either $P \in E$ or P be an accumulation point of E. If $\lim_{x \to P} f(x) = \ell$ and $\lim_{x \to P} f(x) = m$ then $\ell = m$.

PROOF Let $\epsilon > 0$. Choose $\delta_1 > 0$ such that if $0 < |x - P| < \delta_1$ then $|f(x) - \ell| < \epsilon/2$. Similarly choose $\delta_2 > 0$ such that if $0 < |x - P| < \delta_2$ then $|f(x) - m| < \epsilon/2$. Define δ to be the minimum of δ_1 and δ_2. If $0 < |x - P| < \delta$ then the triangle inequality tells us that

$$|\ell - m| = |(\ell - f(x)) + (f(x) - m)|$$
$$\leq |(\ell - f(x)| + |f(x) - m)|$$
$$< \frac{\epsilon}{2} + \frac{\epsilon}{2}$$
$$\lneq \epsilon.$$

Since $|\ell - m| < \epsilon$ for every positive ϵ, we conclude that $\ell = m$. That is the desired result. ∎

The point of the last proposition is that if a limit is calculated by two different methods, then the same answer will result. While of primarily philosophical interest now, this will be important information later.

This is a good time to observe that the limits

$$\lim_{x \to P} f(x)$$

and

$$\lim_{h \to 0} f(P + h)$$

are equal in the sense that if one limit exists then so does the other and they both have the same value.

In order to facilitate checking that certain limits exist, we now record some elementary properties of the limit. This requires that we first recall how functions are combined.

Suppose that f and g are each functions that have domain E. We define the *sum* or *difference* of f and g to be the function

$$(f \pm g)(x) = f(x) \pm g(x),$$

the *product* of f and g to be the function

$$(f \cdot g)(x) = f(x) \cdot g(x),$$

and the quotient of f and g to be

$$\left(\frac{f}{g}\right)(x) = \frac{f(x)}{g(x)}.$$

Notice that the quotient is only defined at points x for which $g(x) \neq 0$. Now we have:

THEOREM 6.2   ELEMENTARY PROPERTIES OF LIMITS OF FUNCTIONS

Let f and g be functions with domain E and fix a point P that is either in E or is an accumulation point of E. Assume that

(i) $\lim_{x \to P} f(x) = \ell$

(ii) $\lim_{x \to P} g(x) = m$.

Then

(a) $\lim_{x \to P}(f \pm g)(x) = \ell \pm m$

(b) $\lim_{x \to P}(f \cdot g)(x) = \ell \cdot m$.

(c) $\lim_{x \to P}(f/g)(x) = \ell/m$ *provided* $m \neq 0$.

PROOF   We prove part (b). Parts (a) and (c) are treated in the Exercises.

Let $\epsilon > 0$. We may also assume that $\epsilon < 1$. Choose $\delta_1 > 0$ such that if $x \in E$ and $0 < |x - P| < \delta_1$ then

$$|f(x) - \ell| < \frac{\epsilon}{2(|m| + 1)}$$

Choose $\delta_2 > 0$ such that if $x \in E$ and $0 < |x - P| < \delta_2$ then

$$|g(x) - m| < \frac{\epsilon}{2(|\ell| + 1)}$$

(Notice that this last inequality implies that $|g(x)| < |m| + |\epsilon|$.) Let δ be the minimum of δ_1 and δ_2. If $x \in E$ and $0 < |x - P| < \delta$ then

$$|f(x) \cdot g(x) - \ell \cdot m| = |(f(x) - \ell) \cdot g(x) + (g(x) - m) \cdot \ell|$$
$$\leq |(f(x) - \ell) \cdot g(x)| + |(g(x) - m) \cdot \ell|$$
$$< \left(\frac{\epsilon}{2(|m| + 1)}\right) \cdot |g(x)| + \left(\frac{\epsilon}{2(|\ell| + 1)}\right) \cdot |\ell|$$
$$\leq \left(\frac{\epsilon}{2(|m| + 1)}\right) \cdot (|\epsilon| + |m|) + \frac{\epsilon}{2}$$
$$\leq \frac{\epsilon}{2} + \frac{\epsilon}{2}$$
$$= \epsilon.$$

Example 6.4

It is a simple matter to check that if $f(x) = x$ then

$$\lim_{x \to P} f(x) = P$$

for every real P. (Indeed, for $\epsilon > 0$ we may take $\delta = \epsilon$.) Also, if $g(x) = \alpha$ is the constant function taking value α then

$$\lim_{x \to P} g(x) = \alpha.$$

It then follows from parts (a) and (b) of the theorem that if $f(x)$ is any polynomial function then

$$\lim_{x \to P} f(x) = f(P).$$

Moreover, if $r(x)$ is any *rational function* (quotient of polynomials) then we may also use part (c) of the theorem to conclude that

$$\lim_{x \to P} r(x) = r(P)$$

for all points P at which the rational function $r(x)$ is defined.   ⬚

Example 6.5

If x is a small positive real number then $0 < \sin x < x$. This is true because $\sin x$ is the nearest distance from the point $(\cos x, \sin x)$ to the x-axis, while x is the distance from that point to the x-axis along an arc. If $\epsilon > 0$ we set $\delta = \epsilon$. We conclude that if $0 < |x - 0| < \delta$ then

$$|\sin x - 0| < |x| < \delta = \epsilon.$$

Since $\sin(-x) = -\sin(-x)$, the same result holds when x is a negative number with small absolute value. Therefore

$$\lim_{x \to 0} \sin x = 0.$$

Since

$$\cos^2 x = 1 - \sin^2 x,$$

we may conclude from the preceding theorem that

$$\lim_{x \to 0} \cos x = 1.$$

Now fix any real number P. We have

$$\lim_{x \to P} \sin x = \lim_{h \to 0} \sin(P + h)$$

$$= \lim_{h \to 0} \sin P \cos h + \cos P \sin h$$

$$= \sin P \cdot 1 + \cos P \cdot 0$$

$$= \sin P.$$

We of course have used parts (a) and (b) of the theorem to commute the limit process with addition and multiplication. A similar argument shows that

$$\lim_{x \to P} \cos x = \cos P. \qquad \qquad \square$$

REMARK 6.3 In the last example we have used the definition of the sine function and the cosine function that you learned in calculus. In Chapter 9, when we learn about series of functions, we will learn a more rigorous method for treating the trigonometric functions. ∎

We conclude by giving a characterization of the limit of a function using sequences.

PROPOSITION 6.4

Let f be a function with domain E and P be either an element of E or an accumulation point of E. Then

$$\lim_{x \to P} f(x) = \ell \qquad \qquad (*)$$

if and only if for any sequence $\{a_j\} \subseteq E \setminus \{P\}$ satisfying $\lim_{j \to \infty} a_j = P$ it holds that

$$\lim_{j \to \infty} f(a_j) = \ell. \qquad \qquad (**)$$

PROOF Assume that condition $(*)$ fails. Then there is an $\epsilon > 0$ such that for no $\delta > 0$ is it the case that when $0 < |x - P| < \delta$ then $|f(x) - \ell| < \epsilon$. Thus for each $\delta = 1/j$ we may choose a number $a_j \in E \setminus \{P\}$ with $0 < |a_j - P| < 1/j$ and $|f(a_j) - \ell| \geq \epsilon$. But then condition $(**)$ fails for this sequence $\{a_j\}$.

If condition $(**)$ fails then there is some sequence $\{a_j\}$ such that $\lim_{j \to \infty} a_j = P$ but $\lim_{j \to \infty} f(a_j) \neq \ell$. This means that there is an $\epsilon > 0$ such that for infinitely many a_j it holds that $|f(a_j) - \ell| \geq \epsilon$. But then no matter how small $\delta > 0$ there will be an a_j satisfying $0 < |a_j - P| < \delta$ (since $a_j \to P$) and $|f(a_j) - \ell| \geq \epsilon$. Thus $(*)$ fails. ∎

6.2   Continuous Functions

DEFINITION 6.6   *Let $E \subseteq \mathbb{R}$ be a set and let f be a real-valued function with domain E. Fix a point $P \in E$. We say that f is **continuous** at P if*

$$\lim_{x \to P} f(x) = f(P).$$

Notice that in the definition of continuity of f we allow P not to be an accumulation point of E. When P is an isolated point of E, then any function is automatically continuous at P. When P is not isolated, there will be several interesting characterizations of continuity at P.

We learned from the last two examples of Section 1 that polynomial functions, $\sin x$, and $\cos x$ are continuous at every real x. A rational function is continuous at every point of its domain.

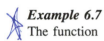

Example 6.7
The function

$$h(x) = \begin{cases} \sin 1/x & \text{if} \quad x \neq 0 \\ 1 & \text{if} \quad x = 0 \end{cases}$$

is discontinuous at 0. The reason is that

$$\lim_{x \to 0} h(x)$$

does not exist. (Details of this assertion are left for you: notice that $h(1/(j\pi)) = 0$ while $h(2/[(4j+1)\pi]) = 1$ for $j = 1, 2, \ldots$.)
The function

$$k(x) = \begin{cases} x \cdot \sin 1/x & \text{if} \quad x \neq 0 \\ 1 & \text{if} \quad x = 0 \end{cases}$$

is also discontinuous at $x = 0$. This time the limit $\lim_{x \to 0} k(x)$ exists (see Example 6.2), but the limit does not agree with $k(0)$.
However the function

$$k(x) = \begin{cases} x \cdot \sin 1/x & \text{if} \quad x \neq 0 \\ 0 & \text{if} \quad x = 0 \end{cases}$$

is continuous at $x = 0$ because the limit at 0 exists and agrees with the value of the function there.   ▯

The arithmetic operations $+, -, \times$, and $\div$ preserve continuity (so long as we avoid division by zero). We now formulate this assertion as a theorem.

THEOREM 6.5
*Let f and g be functions with domain E and let P be a point of E. If f and g
are continuous at P then so are $f \pm g$, $f \cdot g$, and (provided $g(P) \neq 0$) $f \div g$.*

PROOF   Apply Theorem 6.2 of Section 1.   ∎

Continuous functions may also be characterized using sequences:

PROPOSITION 6.6
*Let f be a function with domain E and fix $P \in E$. The function f is continuous
at P if and only if for every sequence $\{a_j\} \subseteq E$ satisfying $\lim_{j \to \infty} a_j = P$ it
holds that*

$$\lim_{j \to \infty} f(a_j) = f(P).$$

PROOF   Apply Proposition 6.4 of Section 1.   ∎

Recall that if g is a function with domain D and range E and if f is a function
with domain E and range H then the *composition* of f and g is

$$f \circ g(x) = f(g(x)).$$

PROPOSITION 6.7
*Let g have domain D and range E and let f have domain E and range H. Let
$P \in D$. Assume that g is continuous at P and that f is continuous at $g(P)$.
Then $f \circ g$ is continuous at P.*

PROOF   Let $\{a_j\}$ be any sequence in D such that $\lim_{j \to \infty} a_j = P$. Then

$$\lim_{j \to \infty} f \circ g(a_j) = \lim_{j \to \infty} f(g(a_j)) = f\left(\lim_{j \to \infty} g(a_j)\right)$$

$$= f\left(g\left(\lim_{j \to \infty} a_j\right)\right) = f(g(P)) = f \circ g(P).$$

Now apply Proposition 6.6.   ∎

REMARK 6.8   It is not the case that if

$$\lim_{x \to P} g(x) = \ell$$

and

$$\lim_{t \to \ell} f(t) = m$$

then

$$\lim_{x \to P} f \circ g(x) = m.$$

A counterexample is given by the functions

$$g(x) = 0$$

$$f(x) = \begin{cases} 2 \text{ if } x \neq 0 \\ 5 \text{ if } x = 0. \end{cases}$$

Notice that $\lim_{x \to 0} g(x) = 0, \lim_{t \to 0} f(x) = 2$, yet $\lim_{x \to 0} f \circ g(x) = 5$.

The additional hypothesis that f be continuous at ℓ is necessary in order to guarantee that the limit of the composition will behave as expected. ∎

Next we explore the topological approach to the concept of continuity. Whereas the analytic approach that we have been discussing so far considers continuity one point at a time, the topological approach considers all points simultaneously. Let us call a function continuous if it is continuous at every point of its domain.

DEFINITION 6.8 *Let f be a function with domain E and let $\mathcal{O}$ be any set of real numbers. We define*

$$f^{-1}(\mathcal{O}) = \{x \in E : f(x) \in \mathcal{O}\}.$$

*We sometimes refer to $f^{-1}(\mathcal{O})$ as the **inverse image** of $\mathcal{O}$ under f.*

THEOREM 6.9
Let f be a function with domain E. The function f is continuous if and only if the inverse image of any open set under f is the intersection of E with an open set.

In particular, if E is open then f is continuous if and only if the inverse image of any open set under f is open.

PROOF Assume that f is continuous. Let $\mathcal{O}$ be any open set and let $P \in f^{-1}(\mathcal{O})$. Then, by definition, $f(P) \in \mathcal{O}$. Since $\mathcal{O}$ is open, there is an $\epsilon > 0$ such that the interval $(f(P) - \epsilon, f(P) + \epsilon)$ lies in $\mathcal{O}$. By the continuity of f we may select a $\delta > 0$ such that if $x \in E$ and $|x - P| < \delta$ then $|f(x) - f(P)| < \epsilon$. In other words, if $x \in E$ and $|x - P| < \delta$ then $f(x) \in \mathcal{O}$ or $x \in f^{-1}(\mathcal{O})$. Thus we have found an open interval $I = (P - \delta, P + \delta)$ about P whose intersection with E is contained in $f^{-1}(\mathcal{O})$. So $f^{-1}(\mathcal{O})$ is the intersection of E with an open set.

Conversely, suppose that for any open set $\mathcal{O} \subseteq \mathbb{R}$ we have that $f^{-1}(\mathcal{O})$ is the intersection of E with an open set. Fix $P \in E$. Choose $\epsilon > 0$. Then the interval $(f(P) - \epsilon, f(P) + \epsilon)$ is an open set. By hypothesis, the set $f^{-1}((f(P) -$

$\epsilon, f(P)+\epsilon))$ is the intersection of E with an open set. This set certainly contains the point P. Thus there is a $\delta > 0$ such that

$$E \cap (P - \delta, P + \delta) \subseteq f^{-1}((f(P) - \epsilon, f(P) + \epsilon)).$$

But that just says that

$$f\left(E \cap (P - \delta, P + \delta)\right) \subseteq (f(P) - \epsilon, f(P) + \epsilon).$$

In other words, if $|x - P| < \delta$ and $x \in E$ then $|f(x) - f(P)| < \epsilon$. But that means that f is continuous at P. ∎

REMARK 6.10   Since any open subset of the real numbers is a countable union of intervals then, in order to check that the inverse image under a function f of every open set is open, it is enough to check that the inverse image of any open interval is open. This is frequently easy to do.

For example, if $f(x) = x^2$ then the inverse image of an open interval (a, b) is $(-\sqrt{b}, -\sqrt{a}) \cup (\sqrt{a}, \sqrt{b})$ if $a > 0$, is $(-\sqrt{b}, \sqrt{b})$ if $a \leq 0, b \geq 0$, and is $\emptyset$ if $a < b < 0$. Thus the function f is continuous.

Note that, by contrast, it is somewhat tedious to give an $\epsilon - \delta$ proof of the continuity of $f(x) = x^2$. ∎

COROLLARY 6.11
Let f be a function with domain E. The function f is continuous if and only if the inverse image of any closed set F under f is the intersection of E with some closed set.

In particular, if E is closed then f is continuous if and only if the inverse image of any closed set F under f is closed.

PROOF   It is enough to prove that

$$f^{-1}\left({}^{c}F\right) = {}^{c}\left(f^{-1}(F)\right).$$

We leave this assertion as an exercise for you. ∎

6.3   Topological Properties and Continuity

Recall that in Chapter 5 we learned a characterization of compact sets in terms of open covers. In Section 2 of the present chapter we learned a characterization of continuous functions in terms of inverse images of open sets. Thus it is not surprising that compact sets and continuous functions interact in a natural way. We explore this interaction in the present section.

DEFINITION 6.9   *Let f be a function with domain E and let G be a subset of E. We define*

$$f(G) = \{f(x) : x \in G\}.$$

*The set $f(G)$ is called the **image** of G under f.*

THEOREM 6.12
The image of a compact set under a continuous function is also compact.

PROOF   Let f be a continuous function with domain E and let K be a subset of E that is compact. Our job is to show that $f(K)$ is compact.

Let $\mathcal{C} = \{\mathcal{O}_\alpha\}$ be an open covering of $f(K)$. Since f is continuous we know that, for each α, the set $f^{-1}(\mathcal{O}_\alpha)$ is the intersection of E with an open set $\mathcal{U}_\alpha$. Let $\hat{\mathcal{C}} = \{\mathcal{U}_\alpha\}_{\alpha \in A}$. Since $\mathcal{C}$ covers $f(K)$ it follows that $\hat{\mathcal{C}}$ covers K. But K is compact; therefore (Theorem 5.14) there is a finite subcovering

$$\{\mathcal{U}_{\alpha_1}, \mathcal{U}_{\alpha_2}, \ldots \mathcal{U}_{\alpha_m}\}$$

of K. But then it follows that $f(\mathcal{U}_{\alpha_1} \cap E), \ldots, f(\mathcal{U}_{\alpha_m} \cap E)$ covers $f(K)$, hence

$$\mathcal{O}_{\alpha_1}, \mathcal{O}_{\alpha_2}, \ldots, \mathcal{O}_{\alpha_m}$$

covers $f(K)$.

We have taken an arbitrary open cover $\mathcal{C}$ for $f(K)$ and extracted from it a finite subcovering. It follows that $f(K)$ is compact.   ∎

It is not the case that the continuous image of a closed set is closed. For instance, take $f(x) = 1/(1 + x^2)$ and $E = \mathbb{R}$: E is closed and f is continuous but $f(E) = (0, 1]$ is not closed.

It is also not the case that the continuous image of a bounded set is bounded. As an example, take $f(x) = 1/x$ and $E = (0, 1)$. Then E is bounded and f continuous but $f(E) = (1, \infty)$ is unbounded.

However, the combined properties of closedness *and* boundedness (that is, compactness) are preserved. That is the content of the preceding theorem.

COROLLARY 6.13
Let f be a function with compact domain K. Then there is a number L such that

$$|f(x)| \leq L$$

for all $x \in K$. In other words, a continuous function on a compact set is bounded.

PROOF   We know from the theorem that $f(K)$ is compact. By Proposition 5.13, we conclude that $f(K)$ is bounded. Thus there is a number L such that $t \leq L$ for all $t \in f(K)$. But that is just the assertion that we wish to prove.   ∎

In fact we can prove an important strengthening of the corollary. Since $f(K)$ is compact, it contains its supremum C and its infimum c. Therefore there must be a number $M \in K$ such that $f(M) = C$ and a number $m \in K$ such that $f(m) = c$. In other words, $f(m) \leq f(x) \leq f(M)$ for all $x \in K$. We summarize:

THEOREM 6.14

Let f be a continuous function on a compact set K. Then there exist numbers m and M in K such that $f(m) \leq f(x) \leq f(M)$ for all $x \in K$. We call m an **absolute minimum** *for f on K and M an* **absolute maximum** *for f on K.*

Notice that in the last theorem M and m need not be unique. For instance, the function $\sin x$ on the compact interval $[0, 4\pi]$ has an absolute minimum at $3\pi/2$ and $7\pi/2$. It has an absolute maximum at $\pi/2$ and at $5\pi/2$.

Now we define a refined type of continuity:

DEFINITION 6.10   *Let f be a function with domain E. We say that f is* **uniformly continuous** *on E if for any $\epsilon > 0$ there is a $\delta > 0$ such that whenever $s, t \in E$ and $|s - t| < \delta$ then $|f(s) - f(t)| < \epsilon$.*

Observe that "uniform continuity" differs from "continuity" in that it treats all points of the domain simultaneously: the $\delta > 0$ that is chosen is independent of the points $s, t \in E$. This difference is highlighted by the next example.

Example 6.11

Consider the function $f(x) = x^2$. Fix a point $P \in \mathbb{R}, P > 0$, and let $\epsilon > 0$. In order to guarantee that $|f(x) - f(P)| < \epsilon$ we must have (for $x > 0$)

$$|x^2 - P^2| < \epsilon$$

or

$$|x - P| < \frac{\epsilon}{x + P}.$$

Since x will range over a neighborhood of P, we see that the required δ in the definition of continuity cannot be larger than $\epsilon/(2P)$. In fact, the choice $|x - P| < \delta = \epsilon/(2P + 1)$ will do the job.

Thus the choice of δ depends not only on ϵ (which we have come to expect) but also on P. In particular, f is not uniformly continuous on $\mathbb{R}$. This is a quantitative reflection of the fact that the graph of f becomes ever steeper as the variable moves to the right.

Notice that the same calculation shows that the function $f(x) = x^2$ with domain $[a, b], 0 < a < b < \infty$, *is* uniformly continuous. □

The main result about uniform continuity now follows:

THEOREM 6.15

Let f be a continuous function with compact domain K. Then f is uniformly continuous on K.

PROOF   Pick $\epsilon > 0$. By the definition of continuity there is for each point $x \in K$ a number $\delta_x > 0$ such that if $|x - t| < \delta_x$ then $|f(t) - f(x)| < \epsilon/2$. The intervals $I_x = (x - \delta_x/2, x + \delta_x/2)$ form an open covering of K. Since K is compact, we may therefore (by 5.14) extract a finite subcovering

$$I_{x_1}, \ldots I_{x_m}.$$

Now let $\delta = \min\{\delta_{x_1}/2, \ldots, \delta_{x_m}/2\} > 0$. If $s, t \in K$ and $|s - t| < \delta$, then $s \in I_{x_j}$ for some $1 \le j \le m$. It follows that

$$|s - x_j| < \delta_{x_j}/2$$

and

$$|t - x_j| \le |t - s| + |s - x_j| < \delta + \delta_{x_j}/2 \le \delta_{x_j}/2 + \delta_{x_j}/2 = \delta_{x_j}.$$

We know that

$$|f(s) - f(t)| \le |f(s) - f(x_j)| + |f(x_j) - f(t)|.$$

But since each of s and t is within δ_{x_j} of x_j, we may conclude that the last line is less than

$$\frac{\epsilon}{2} + \frac{\epsilon}{2} = \epsilon.$$

Notice that our choice of δ does not depend on s and t (indeed, we chose δ *before* we chose s and t). We conclude that f is uniformly continuous. ∎

REMARK 6.16   Where in the proof did the compactness play a role? We defined δ to be the minimum of $\delta_{x_1}, \ldots \delta_{x_m}$. In order to guarantee that δ be *positive*, it is crucial that we take the minimum of *finitely many* positive numbers. ∎

Example 6.12

The function $f(x) = \sin(1/x)$ is continuous on the domain $E = (0, \infty)$ since it is the composition of continuous functions. However it is not uniformly continuous since

$$\left| f\left(\frac{1}{2j\pi}\right) - f\left(\frac{1}{\frac{(4j+1)\pi}{2}}\right) \right| = 1$$

for $j = 1, 2, \ldots$. Thus, even though the arguments are becoming arbitrarily close together, the images of these arguments remain bounded apart. We conclude that f cannot be uniformly continuous.

However, if f is considered as a function on any interval of the form $[a, b]$, $0 < a < b < \infty$, then the preceding theorem tells us that f is uniformly continuous. ☐

As an exercise, you should check that

$$g(x) = \begin{cases} x\sin(1/x) & \text{if} \quad x \neq 0 \\ 0 & \text{if} \quad x = 0 \end{cases}$$

is uniformly continuous on any interval of the form $[-N, N]$.

Next we show that continuous functions preserve connectedness.

THEOREM 6.17

Let f be a continuous function with domain an open interval I. Suppose that L is a connected subset of I. Then $f(L)$ is connected.

PROOF Suppose to the contrary that there are open sets U and V such that

$$U \cap f(L) \neq \emptyset \;, \quad V \cap f(L) \neq \emptyset,$$

$$\big(U \cap f(L)\big) \cap \big(V \cap f(L)\big) = \emptyset,$$

and

$$f(L) = \big(U \cap f(L)\big) \cup \big(V \cap f(L)\big).$$

Since f is continuous, $f^{-1}(U)$ and $f^{-1}(V)$ are open. They each have nonempty intersection with L since $U \cap f(L)$ and $V \cap f(L)$ are nonempty. By the definition of f^{-1}, they are certainly disjoint. And since $U \cup V$ contains $f(L)$, it follows by definition that $f^{-1}(U) \cup f^{-1}(V)$ contains L. But this shows that L is disconnected, and that is a contradiction. ∎

COROLLARY 6.18   THE INTERMEDIATE VALUE THEOREM
Let f be a continuous function whose domain contain the interval $[a, b]$. Let γ be a number that lies between $f(a)$ and $f(b)$. Then there is a number c between a and b such that $f(c) = \gamma$.

PROOF   The set $[a, b]$ is connected. Therefore $f([a, b])$ is connected. But $f([a, b])$ contains the points $f(a)$ and $f(b)$. By connectivity, $f([a, b])$ must contain the interval that has $f(a)$ and $f(b)$ as endpoints. In particular, $f([a, b])$ must contain any number γ that lies between $f(a)$ and $f(b)$. But this just says that there is a number c lying between a and b such that $f(c) = \gamma$.   ∎

6.4   Classifying Discontinuities and Monotonicity

We begin by refining our notion of limit:

DEFINITION 6.13   *Fix $P \in \mathbb{R}$. Let f be a function with domain E. Fix a point $P \in E$. We say that f has **left limit** ℓ at P, and write*

$$\lim_{x \to P-} f(x) = \ell,$$

if for every $\epsilon > 0$ there is a $\delta > 0$ such that whenever $P - \delta < x < P$ and $x \in E$ then it holds that

$$|f(x) - \ell| < \epsilon.$$

*We say that f has **right limit** m at P, and write*

$$\lim_{x \to P+} f(x) = m,$$

if for every $\epsilon > 0$ there is a $\delta > 0$ such that whenever $P < x < P + \delta$ and $x \in E$ then it holds that

$$|f(x) - m| < \epsilon.$$

This definition simply formalizes the notion of either letting x tend to P from the left only or from the right only.

Let f be a function with domain E. Let P in E and assume that f is discontinuous at P. There are two ways in which this discontinuity can occur:

I. If $\lim_{x \to P-} f(x)$ and $\lim_{x \to P+} f(x)$ both exist but either do not equal each other or do not equal $f(P)$ then we say that f has a *discontinuity of the first kind* (or sometimes a *simple discontinuity*) at P.

II. If either $\lim_{x \to P-}$ does not exist or $\lim_{x \to P+}$ does not exist then we say that f has a *discontinuity of the second kind* at P.

f_ not def. @ 0.
2nd lim_{p+} or @ p^- DNE

Example 6.14
Define

$$f(x) = \begin{cases} \sin(1/x) & \text{if } x \neq 0 \\ 0 & \text{if } x = 0 \end{cases}$$

$$g(x) = \begin{cases} 1 & \text{if } x > 0 \\ 0 & \text{if } x = 0 \\ -1 & \text{if } x < 0 \end{cases}$$

$$h(x) = \begin{cases} 1 & \text{if } x \text{ is irrational} \\ 0 & \text{if } x \text{ is rational.} \end{cases}$$

Then f has a discontinuity of the second kind at 0 while g has a discontinuity of the first kind at 0. The function h has a discontinuity of the second kind at every point. ⬜

DEFINITION 6.15 *Let f be a function whose domain contains an open interval (a, b). We say that f is **monotonically increasing** on (a, b) if whenever $a < s < t < b$ it holds that $f(s) \leq f(t)$. We say that f is **monotonically decreasing** on (a, b) if whenever $a < s < t < b$ it holds that $f(s) \geq f(t)$.*

Functions that are either monotonically increasing or monotonically decreasing are simply referred to as "monotonic."

As with sequences, the word "monotonic" is superfluous in many contexts. But its use is traditional and occasionally convenient.

PROPOSITION 6.19
Let f be a monotonic function on an open interval (a, b). Then all of the discontinuities of f are of the first kind.

PROOF It is enough to show that for each $P \in (a, b)$ the limits

$$\lim_{x \to P-} f(x)$$

and

$$\lim_{x \to P+} f(x)$$

exist.

Let us first assume that f is monotonically increasing. Fix $P \in (a, b)$. If $a < s < P$ then $f(s) \leq f(P)$. Therefore $S = \{f(s) : a < s < P\}$ is bounded above. Let M be the least upper bound of S. Pick $\epsilon > 0$. By definition of least upper bound there must be an $f(s) \in S$ such that $f(s) - M < \epsilon$. Let $\delta = |P - s|$. If $P - \delta < t < P$ then $s < t < P$ and $f(s) \leq f(t) \leq M$ so $|f(t) - M| < \epsilon$. Thus $\lim_{x \to P-} f(x)$ exists and equals M.

If we set m equal to the infimum of the set $T = \{f(t) : P < t < b\}$ then a similar argument shows that $\lim_{x \to P+} f(x)$ exists and equals m. That completes the proof. ∎

COROLLARY 6.20

Let f be a monotonic function on an interval (a, b). Then f has at most countably many discontinuities.

PROOF   Assume for simplicity that f is monotonically increasing. If P is a discontinuity then the proposition tells us that

$$\lim_{x \to P-} f(x) < \lim_{x \to P+} f(x).$$

Therefore there is a rational number q_P between $\lim_{x \to P-} f(x)$ and $\lim_{x \to P+} f(x)$. Notice that different discontinuities will have different rational numbers associated to them because if $\hat{P}$ is another discontinuity and, say, $\hat{P} < P$ then

$$\lim_{x \to \hat{P}-} f(x) < q_{\hat{P}} < \lim_{x \to \hat{P}+} f(x) \leq \lim_{x \to P-} f(x) < q_P < \lim_{x \to P+} f(x).$$

Thus we have exhibited a one-to-one function from the set of discontinuities of f into the set of rational numbers. It follows that the set of discontinuities is countable. ∎

A continuous function f has the property that the inverse image under f of any open set is open. However, it is not generally true that the *image* under f itself of any open set is open. A counterexample is the function $f(x) = x^2$ and the open set $\mathcal{O} = (-1, 1)$ whose image under f is $[0, 1)$. However with some additional hypotheses it is the case that continuous functions take open sets to open sets:

THEOREM 6.21

Let f be a continuous function whose domain is a compact set K. Let $\mathcal{O}$ be any open set in $\mathbb{R}$. Then $f(K \cap \mathcal{O})$ has the form $f(K) \cap \mathcal{U}$ for some open set $\mathcal{U} \subseteq \mathbb{R}$.

PROOF   Let $E = K \setminus \mathcal{O}$. Then E is closed (because K is) and bounded (because K is). Thus E is compact. By Theorem 6.12, $f(E)$ must be compact. In particular, it is closed. Let $\mathcal{U} = \mathbb{R} \setminus f(E)$. Then $\mathcal{U}$ is open and $f(K \cap \mathcal{O}) = f(K) \cap \mathcal{U}$. That is the desired result. ∎

Suppose that f is a function on (a, b) such that $a < s < t < b$ implies $f(s) < f(t)$. Such a function is called *strictly monotonically increasing* (*strictly monotonically decreasing* functions are defined similarly). It is clear that a strictly monotonically increasing (resp. decreasing) function is one-to-one, hence has an inverse. Now we prove:

THEOREM 6.22
Let f be a strictly monotone, continuous function with domain $[a, b]$. Then f^{-1} exists and is continuous.

PROOF Assume without loss of generality that f is strictly monotone *increasing*. Let us extend f to the entire real line by defining

$$f(x) = \begin{cases} (x - a) + f(a) & \text{if} \quad x < a \\ \text{as given} & \text{if} \quad a \leq x \leq b \\ (x - b) + f(b) & \text{if} \quad x > b. \end{cases}$$

Then it is easy to see that this extended version of f is still continuous and is strictly monotone increasing on all of $\mathbb{R}$.

That f^{-1} exists has already been discussed. The extended function f takes any open interval (c, d) to the open interval $(f(c), f(d))$. Since any open set is a union of open intervals, we see that f takes any open set to an open set. In other words, $\left[f^{-1}\right]^{-1}$ takes open sets to open sets. But this just says that f^{-1} is continuous.

Since the inverse of the extended function f is continuous, then so is the inverse of the original function f. That completes the proof. ∎

Exercises

6.1 Let f and g be functions on a set $A = (a, c) \cup (c, b)$ and assume that $f(x) \leq g(x)$ for all $x \in A$. Assuming that both limits exist, show that

$$\lim_{x \to c} f(x) \leq \lim_{x \to c} g(x).$$

Does the conclusion improve if we assume that $f(x) < g(x)$ for all $x \in A$?

6.2 If f is defined on a set $A = (a, c) \cup (c, b)$ and if $\lim_{x \to c} f(x) = r > 0$ then prove that there is a $\delta > 0$ such that if $0 < |x - c| < \delta$ then $|f(x)| > r/2$.

6.3 Give an example of a function f for which the situation in Exercise 2 obtains but such that f is not continuous at the point c.

6.4 Give an example of a continuous function f and a connected set E such that $f^{-1}(E)$ is not connected.

6.5 Give an example of a continuous function f and a compact set K such that $f^{-1}(K)$ is not a compact set.

6.6 Let A be any countable subset of the reals. Construct a monotone increasing function whose set of points of discontinuity is precisely the set A.

6.7 A function f with domain E said to satisfy a *Lipschitz condition* of order α, $0 < \alpha \leq 1$, if there is a constant $C > 0$ such that for any $s, t \in E$ it holds that $|f(s) - f(t)| \leq C \cdot |s - t|^{\alpha}$. Prove that such a function must be uniformly continuous.

6.8 Let S be any subset of $\mathbb{R}$. Define the function
$$f(x) = \inf\{|x - s| : s \in S\}.$$
Prove that f is uniformly continuous.

6.9 Define the function
$$g(x) = \begin{cases} 0 & \text{if } x \text{ is irrational} \\ x & \text{if } x \text{ is rational.} \end{cases}$$
At which points x is g continuous? At which points is it discontinuous?

6.10 Define the function $g(x)$ to take the value 0 at irrational values of x and to take the value $1/q$ when $x = p/q$ is a rational number in lowest terms, $q > 0$. At which points is g continuous? At which points is the function discontinuous?

6.11 Let f be any function whose domain is the entire real line. If A and B are disjoint sets does it follow that $f(A)$ and $f(B)$ are disjoint sets? If C and D are disjoint sets does it follow that $f^{-1}(C)$ and $f^{-1}(D)$ are disjoint?

6.12 Let f be any function whose domain is the entire real line. If A and B are sets then is $f(A \cup B) = f(A) \cup f(B)$? If C and D are sets then is $f^{-1}(C \cup D) = f^{-1}(C) \cup f^{-1}(D)$? What is the answer to these questions if we replace $\cup$ by $\cap$?

6.13 Give an example of two functions, discontinuous at $x = 0$, whose sum *is* continuous at $x = 0$. Give an example of two such functions whose product is continuous at $x = 0$.

6.14 Let f be a function with domain the real numbers. If $f^2(x) = f(x) \cdot f(x)$ is continuous, does it follow that f is continuous? If $f^3(x) = f(x) \cdot f(x) \cdot f(x)$ is continuous, does it follow that f is continuous?

6.15 Fix an interval (a, b). Is the collection of monotone increasing functions on (a, b) closed under $+, -, \times$, or $\div$?

6.16 True or false: If f is a function with domain and range the real numbers and is both one-to-one and onto, then f must be either monotone increasing or monotone decreasing. Does your answer change if we assume that f is continuous?

6.17 Prove that the function $f(x) = \sin x$ can be written, on the interval $(0, 4\pi)$, as the difference of two monotone increasing functions.

6.18 In Remark 6.8 we suggested a generalization of Proposition 6.7. Prove this generalization. (Hint: g need not be continuous at P.)

6.19 Let f be a continuous function whose domain contains a closed, bounded interval $[a, b]$. What topological properties does $f([a, b])$ possess? Is this set necessarily an interval?

6.20 A function f from an interval (a, b) to an interval (c, d) is called "proper" if for any compact set $K \subseteq (c, d)$ it holds that $f^{-1}(K)$ is compact. Prove that if f is proper then either
$$\lim_{x \to a^+} f(x) = c \quad \text{or} \quad \lim_{x \to a^+} f(x) = d.$$
Likewise prove that either
$$\lim_{x \to b^-} f(x) = c \quad \text{or} \quad \lim_{x \to b^-} f(x) = d.$$

6.21 We know that the continuous image of a connected set (i.e., an interval) is also a connected set (another interval). Suppose now that A is the union of k disjoint intervals and that f is a continuous function. What can you say about the set $f(A)$?

6.22 A function f with domain A and range B is called a *homeomorphism* if it is one-to-one, onto, continuous, and has a continuous inverse. If such an f exists then we say that A and B are *homeomorphic*. Which sets of reals are homeomorphic to the open unit interval $(0,1)$? Which sets of reals are homeomorphic to the closed unit interval $[0,1]$?

6.23 Let f be a continuous function with domain $[0,1]$ and range $[0,1]$. Prove that there exists a point $P \in [0,1]$ such that $f(P) = P$. (Hint: Apply the Intermediate Value Theorem to the function $g(x) = f(x) - x$.) Prove that this result is false if the domain and range of the function are both $(0,1)$.

6.24 Refer to Exercise 22 for terminology. Show that there is no homeomorphism from the real line to the interval $[0,1)$.

6.25 Is the composition of uniformly continuous functions uniformly continuous?

6.26 Let f be a continuous function and let $\{a_j\}$ be a Cauchy sequence in the domain of f. Does it follow that $\{f(a_j)\}$ is a Cauchy sequence? What if we assume that f is uniformly continuous?

6.27 Let E be any closed set of real numbers. Prove that there is a continuous function f with domain $\mathbb{R}$ such that $\{x : f(x) = 0\} = E$.

6.28 Let E and F be disjoint closed sets of real numbers. Prove that there is a continuous function f with domain the real numbers such that $\{x : f(x) = 0\} = E$ and $\{x : f(x) = 1\} = F$.

6.29 If K and L are sets then define
$$K + L = \{k + l : k \in K \quad \text{and} \quad l \in L\}.$$
If K and L are compact then prove that $K + L$ is compact. If K and L are merely closed, does it follow that $K + L$ is closed?

6.30 Let f be a function with domain $\mathbb{R}$. Prove that the set of discontinuities of the first kind for f is countable. (Hint: If the left and right limits at a point disagree then you can slip a rational number between them, but the same left and right limits can occur at different points of the domain so you must use rational numbers to keep track of them as well.)

6.31 Prove parts (a) and (c) of Theorem 6.2.

6.32 Let f be a continuous function whose domain contains an open interval (a,b). What form can $f(a,b)$ have? (Hint: There are just four possibilities.)

6.33 Let $I \subseteq \mathbb{R}$ be an open interval and $f : I \to \mathbb{R}$ a function. We say that f is *convex* if whenever $\alpha, \beta \in I$ and $0 \leq t \leq 1$ then
$$f((1 - t)\alpha + t\beta) \leq (1 - t)f(\alpha) + tf(\beta).$$
Prove that a convex function must be continuous. What does this definition of convex function have to do with the notion of "concave up" that you learned in calculus?

6.34 Explain why it is inappropriate to calculate the limit of a function f on a set E at a point P when P is in the interior of the complement of E.

7

Differentiation of Functions

7.1   The Concept of Derivative

Let f be a function with domain an open interval I. If $x \in I$ then the quantity

$$\frac{f(t) - f(x)}{t - x}$$

measures the slope of the chord of the graph of f that connects the points $(x, f(x))$ and $(t, f(t))$. If we let $t \to x$ then the limit of the quantity represented by this "Newton quotient" should represent the slope of the graph *at the point* x. These considerations motivate the definition of the derivative:

DEFINITION 7.1   *If f is a function with domain an open interval I and if $x \in I$ then the limit*

$$\lim_{t \to x} \frac{f(t) - f(x)}{t - x},$$

*when it exists, is called the **derivative** of f at x. If the derivative of f at x exists then we say that f is **differentiable** at x. If f is differentiable at every $x \in I$ then we say that f is **differentiable on** I.*

We write the derivative of f at x either as

$$f'(x) \quad or \quad \frac{d}{dx}f \quad or \quad \frac{df}{dx}.$$

We begin our discussion of the derivative by establishing some basic properties and relating the notion of derivative to continuity.

LEMMA 7.1

If f is differentiable at a point x then f is continuous at x. In particular, $\lim_{t \to x} f(t) = f(x)$.

PROOF   We use Theorem 6.2(b) about limits to see that

$$\lim_{t \to x} \left(f(t) - f(x) \right) = \lim_{t \to x} \left((t - x) \cdot \frac{f(t) - f(x)}{t - x} \right)$$

$$= \lim_{t \to x} (t - x) \cdot \lim_{t \to x} \frac{f(t) - f(x)}{t - x}$$

$$= 0 \cdot f'(x)$$

$$= 0.$$

Therefore $\lim_{t \to x} f(t) = f(x)$ and f is continuous at x.   ∎

Thus all differentiable functions are continuous: differentiability is a stronger property than continuity.

THEOREM 7.2

Assume that f and g are functions with domain an open interval I and that f and g are differentiable at $x \in I$. Then $f \pm g, f \cdot g$, and f/g are differentiable at x (for f/g we assume that $g(x) \neq 0$.) Moreover,

(a) $(f \pm g)'(x) = f'(x) \pm g'(x)$;

(b) $(f \cdot g)'(x) = f'(x) \cdot g(x) + f(x) \cdot g'(x)$;

(c) $\left(\frac{f}{g} \right)'(x) = \frac{g(x) \cdot f'(x) - f(x) \cdot g'(x)}{g^2(x)}$.

PROOF   Assertion (a) is easy and we leave it as an exercise for you.
   For (b), we write

$$\lim_{t \to x} \frac{(f \cdot g)(t) - (f \cdot g)(x)}{t - x} = \lim_{t \to x} \left(\frac{(f(t) - f(x)) \cdot g(t)}{t - x} \right.$$

$$\left. + \frac{(g(t) - g(x)) \cdot f(x)}{t - x} \right)$$

$$= \lim_{t \to x} \left(\frac{(f(t) - f(x)) \cdot g(t)}{t - x} \right)$$

$$+ \lim_{t \to x} \left(\frac{(g(t) - g(x)) \cdot f(x)}{t - x} \right)$$

$$= \lim_{t \to x} \left(\frac{(f(t) - f(x))}{t - x} \right) \cdot \left(\lim_{t \to x} g(t) \right)$$

$$+ \lim_{t \to x} \left(\frac{(g(t) - g(x))}{t - x} \right) \cdot \left(\lim_{t \to x} f(x) \right),$$

where we have used Theorem 6.2 about limits. Now the first limit is the derivative of f at x, while the third limit is the derivative of g at x. Also notice that the limit of $g(t)$ equals $g(x)$ by the lemma. The result is that the last line equals

$$f'(x) \cdot g(x) + g'(x) \cdot f(x),$$

as desired.

To prove (c), write

$$\lim_{t \to x} \frac{(f/g)(t) - (f/g)(x)}{t - x} = \lim_{t \to x} \frac{1}{g(t) \cdot g(x)} \left(\frac{f(t) - f(x)}{t - x} \cdot g(x) \right.$$

$$\left. - \frac{g(t) - g(x)}{t - x} \cdot f(x) \right)$$

The proof is now completed by using Theorem 6.2 about limits to evaluate the individual limits in this expression. ∎

That $f(x) = x$ is differentiable follows from

$$\lim_{t \to x} \frac{t - x}{t - x} = 1.$$

Any constant function is differentiable (with derivative identically zero) by a similar argument. It follows from the theorem that any polynomial function is differentiable.

On the other hand, the function $f(x) = |x|$ is *not* differentiable at the point $x = 0$. This is so because

$$\lim_{t \to 0^-} \frac{|t| - |0|}{t - x} = \lim_{t \to 0^-} \frac{-t - 0}{t - 0} = -1$$

while

$$\lim_{t \to 0^+} \frac{|t| - |0|}{t - x} = \lim_{t \to 0^+} \frac{t - 0}{t - 0} = 1.$$

So the required limit does not exist.

Since the subject of differential calculus is concerned with learning uses of the derivative, it concentrates on functions which *are* differentiable. One comes away from the subject with the impression that most functions are differentiable except at a few isolated points — as is the case with the function $f(x) = |x|$. Indeed, this was what the mathematicians of the nineteenth century thought. It therefore came as a shock when Karl Weierstrass produced a continuous function that is not differentiable at *any point*. In a sense that will be made precise in Chapter 12, *most* continuous functions are of this nature: their graphs "wiggle" so much that they cannot have a tangent line at any point. Now we turn to a variation of the example of Weierstrass that is due to B. L. van der Waerden (1903–).

THEOREM 7.3

Define a function ψ with domain $\mathbb{R}$ by the rule

$$\psi(x) = \begin{cases} x - n & \text{if } n \leq x < n+1 \text{ and } n \text{ is even} \\ n+1-x & \text{if } n \leq x < n+1 \text{ and } n \text{ is odd.} \end{cases}$$

The graph of this function is exhibited in the figure:

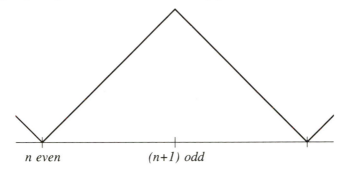

n even   *(n+1) odd*

Then the function

$$f(x) = \sum_{j=1}^{\infty} \left(\frac{3}{4}\right)^j \psi\left(4^j x\right)$$

is continuous at every real x and differentiable at no real x.

PROOF Since we have not yet discussed series of functions, we take a moment to understand the definition of f. Fix a real x. Then the series becomes a series of numbers, and the jth summand does not exceed $(3/4)^j$ in absolute value. Thus the series converges absolutely; therefore it converges. So it is clear that the displayed formula defines a function of x.

Step I: f is continuous

To see that f is continuous, pick an $\epsilon > 0$. Choose N so large that

$$\sum_{j=N+1}^{\infty} \left(\frac{3}{4}\right)^j < \frac{\epsilon}{4}.$$

(We can of course do this because the series $\sum (3/4)^j$ converges.) Now fix x. Observe that since ψ is continuous and the graph of ψ is composed of segments of slope 1 we have

$$|\psi(s) - \psi(t)| \leq |s - t|$$

for all s and t. Moreover, $|\psi(s) - \psi(t)| \leq 1$ for all s, t.

For $j = 1, 2, \ldots, N$ pick $\delta_j > 0$ so that when $|x - t| < \delta_j$ then

$$\left| \psi \left(4^j t \right) - \psi \left(4^j x \right) \right| < \frac{\epsilon}{8}.$$

Let δ be the minimum of $\delta_1, \ldots, \delta_N$.

Now if $|t - x| < \delta$ then

$$|f(t) - f(x)| = \left| \sum_{j=1}^{N} \left(\frac{3}{4} \right)^j \cdot \left(\psi(4^j t) - \psi(4^j x) \right) \right.$$

$$\left. + \sum_{j=N+1}^{\infty} \left(\frac{3}{4} \right)^j \cdot \left(\psi(4^j t) - \psi(4^j x) \right) \right|$$

$$\leq \sum_{j=1}^{N} \left(\frac{3}{4} \right)^j \left| \left(\psi(4^j t) - \psi(4^j x) \right) \right|$$

$$+ \sum_{j=N+1}^{\infty} \left(\frac{3}{4} \right)^j \left| \psi(4^j t) - \psi(4^j x) \right|$$

$$\leq \sum_{j=1}^{N} \left(\frac{3}{4} \right)^j \cdot \frac{\epsilon}{8} + \sum_{j=N+1}^{\infty} \left(\frac{3}{4} \right)^j.$$

Here we have used the choice of δ to estimate the summands in the first sum. The first sum is thus less than $\epsilon/2$ (just notice that $\sum_{j=1}^{\infty} (3/4)^j < 4$). The second sum is less than $\epsilon/4$ by the choice of N. Altogether then

$$|f(t) - f(x)| < \epsilon$$

whenever $|t - x| < \delta$. Therefore f is continuous, indeed uniformly so.

Step II: f is nowhere differentiable

Fix x. For $\ell = 1, 2, \ldots$ define $t_\ell = x \pm 4^{-\ell}/2$. We will say whether the sign is plus or minus in a moment (this will depend on the position of x relative to the integers). Then

$$\left| \frac{f(t_\ell) - f(x)}{t_\ell - x} \right| = \left| \frac{1}{t_\ell - x} \left[\sum_{j=1}^{\ell} \left(\frac{3}{4} \right)^j \left(\psi(4^j t_\ell) - \psi(4^j x) \right) \right. \right.$$

$$\left. \left. + \sum_{j=\ell+1}^{\infty} \left(\frac{3}{4} \right)^j \left(\psi(4^j t_\ell) - \psi(4^j x) \right) \right] \right|. \qquad (*)$$

Notice that when $j \geq \ell + 1$ then $4^j t_\ell$ and $4^j x$ differ by an even integer. Since ψ has period 2, we find that each of the summands in the second sum is 0. Next we turn to the first sum.

We choose the sign — plus or minus — in the definition of t_ℓ so that there is no integer lying between $4^\ell t_\ell$ and $4^\ell x$. We can do this because the two numbers differ by $1/2$. Then the ℓth summand has magnitude

$$(3/4)^\ell \cdot |4^\ell t_\ell - 4^\ell x| = 3^\ell |t_\ell - x|.$$

On the other hand, the first $\ell - 1$ summands add up to not more than

$$\sum_{j=1}^{\ell-1} \left(\frac{3}{4}\right)^j \cdot |4^j t_\ell - 4^j x| = \sum_{j=1}^{\ell-1} 3^j \cdot 4^{-\ell}/2 \leq \frac{3^\ell - 1}{3 - 1} \cdot 4^{-\ell}/2 \leq 3^\ell \cdot 4^{-\ell-1}.$$

It follows that

$$\left| \frac{f(t_\ell) - f(x)}{t_\ell - x} \right| = \frac{1}{|t_\ell - x|} \cdot \left| \sum_{j=1}^{\ell} \left(\frac{3}{4}\right)^j \left(\psi(4^j t_\ell) - \psi(4^j x)\right) \right|$$

$$= \frac{1}{|t_\ell - x|} \cdot \left| \sum_{j=1}^{\ell-1} \left(\frac{3}{4}\right)^j \left(\psi(4^j t_\ell) - \psi(4^j x)\right) \right.$$

$$\left. + \left(\frac{3}{4}\right)^\ell \left(\psi(4^\ell t_\ell) - \psi(4^\ell x)\right) \right|$$

$$\geq \frac{1}{|t_\ell - x|} \cdot \left| \left(\frac{3}{4}\right)^\ell \psi(4^\ell t_\ell) - \left(\frac{3}{4}\right)^\ell \psi(4^\ell x) \right|$$

$$- \frac{1}{|t_\ell - x|} \left| \sum_{j=1}^{\ell-1} \left(\frac{3}{4}\right)^j \left(\psi(4^j t_\ell) - \psi(4^j x)\right) \right|$$

$$\geq 3^\ell - \frac{1}{(4^{-\ell}/2)} \cdot 3^\ell \cdot 4^{-\ell-1}$$

$$\geq 3^{\ell-1}.$$

Thus $t_\ell \to x$ but the Newton quotients blow up. Therefore the limit

$$\lim_{t \to x} \frac{f(t) - f(x)}{t - x}$$

cannot exist. The function f is not differentiable at x. ∎

The proof of the last theorem was long, but the idea is simple: the function f is built by piling oscillations on top of oscillations. When the ℓth oscillation is added, it is made very small in size so that it does not cancel the previous oscillations. But it is made very steep so that it will cause the derivative to become large.

The practical meaning of Weierstrass's example is that we should realize that differentiability is a very strong and special property of functions. Most continuous functions are not differentiable at any point. When we are proving theorems about continuous functions, we should not think of them in terms of properties of differentiable functions.

Next we turn to the Chain Rule.

THEOREM 7.4
Let g be a differentiable function on an open interval I and let f be a differentiable function on an open interval that contains the range of g. Then $f \circ g$ is differentiable on the interval I and

$$(f \circ g)'(x) = f'(g(x)) \cdot g'(x)$$

for each $x \in I$.

PROOF   We use the notation Δt to stand for an increment in the variable t. Let us use the symbol $\mathcal{V}(r)$ to stand for any expression that tends to 0 as $\Delta r \to 0$. Fix $x \in I$. Set $r = g(x)$. By hypothesis,

$$\lim_{\Delta r \to 0} \frac{f(r + \Delta r) - f(r)}{\Delta r} = f'(r)$$

or

$$\frac{f(r + \Delta r) - f(r)}{\Delta r} - f'(r) = \mathcal{V}(r)$$

or

$$f(r + \Delta r) = f(r) + \Delta r \cdot f'(r) + \Delta r \cdot \mathcal{V}(r). \tag{$*$}$$

Notice that equation $(*)$ is valid even when $\Delta r = 0$. Since Δr in equation $(*)$ can be any small quantity, we set

$$\Delta r = \Delta x \cdot [g'(x) + \mathcal{V}(x)].$$

Substituting this expression into $(*)$ and using the fact that $r = g(x)$ yields that

$$f(g(x) + \Delta x[g'(x) + \mathcal{V}(x)]) = f(r) + \big(\Delta x \cdot [g'(x) + \mathcal{V}(x)]\big) \cdot f'(r)$$
$$+ \big(\Delta x \cdot [g'(x) + \mathcal{V}(x)]\big) \cdot \mathcal{V}(r)$$
$$= f(g(x)) + \Delta x \cdot f'(g(x)) \cdot g'(x)$$
$$+ \Delta x \cdot \mathcal{V}(x). \tag{$**$}$$

Just as we derived $(*)$, we may also obtain

$$g(x + \Delta x) = g(x) + \Delta x \cdot g'(x) + \Delta x \cdot \mathcal{V}(x).$$

We may substitute this into the left side of $(**)$ to obtain

$$f(g(x + \Delta x)) = f(g(x)) + \Delta x \cdot f'(g(x)) \cdot g'(x) + \Delta x \cdot \mathcal{V}(x).$$

With some algebra this can be rewritten as

$$\frac{f(g(x + \Delta x)) - f(g(x))}{\Delta x} - f'(g(x)) \cdot g'(x) = \mathcal{V}(x).$$

But this just says that

$$\lim_{\Delta x \to 0} \frac{(f \circ g)(x + \Delta x) - (f \circ g)(x)}{\Delta x} = f'(g(x)) \cdot g'(x).$$

That is, $(f \circ g)'(x)$ exists and equals $f'(g(x)) \cdot g'(x)$, as desired.   ∎

7.2   The Mean Value Theorem and Applications

We begin this section with some remarks about local maxima and minima of functions.

DEFINITION 7.2   *Let f be a function with domain (a, b). A point $x \in (a, b)$ is called a **local maximum** for f if there is a $\delta > 0$ such that $f(t) \leq f(x)$ for all $t \in (x - \delta, x + \delta)$. A point $x \in (a, b)$ is called a **local minimum** for f if there is a $\delta > 0$ such that $f(t) \geq f(x)$ for all $t \in (x - \delta, x + \delta)$.*

*Local minima (plural of minimum) and local maxima (plural of maximum) are referred to collectively as **local extrema**.*

PROPOSITION 7.5
If f is a function with domain (a, b), if f has a local extremum at $x \in (a, b)$, and if f is differentiable at x, then $f'(x) = 0$.

PROOF   Suppose that x is a local minimum. Then there is an $\epsilon > 0$ such that if $-\epsilon < t < x$ then $f(t) \geq f(x)$. Then

$$\frac{f(t) - f(x)}{t - x} \leq 0.$$

Letting $t \to x$, it follows that $f'(x) \leq 0$. Similarly, if $x < t < x + \epsilon$ then

$$\frac{f(t) - f(x)}{t - x} \geq 0.$$

It follows that $f'(x) \geq 0$. We must conclude that $f'(x) = 0$.

A similar argument applies if x is a local maximum. The proof is complete.   ∎

Before going on to mean value theorems, we provide a striking application of the proposition:

THEOREM 7.6   DARBOUX'S THEOREM

Let f be a differentiable function on an open interval I. Pick points $s < t$ in I and suppose that $f'(s) < \rho < f'(t)$. Then there is a point u between s and t such that $f'(u) = \rho$.

PROOF   Consider the function $g(x) = f(x) - \rho x$. Then $g'(s) < 0$ and $g'(t) > 0$. Assume for simplicity that $s < t$. The sign of the derivative at s guarantees that $g(\hat{s}) < g(s)$ for $\hat{s}$ greater than s and near s. The sign of the derivative at t guarantees that $g(\hat{t}) < g(t)$ for $\hat{t}$ less than t and near t. Thus the minimum of the continuous function g on the compact interval $[s, t]$ must occur at some point u in the interior (s, t). The proposition guarantees that $g'(u) = 0$, or $f'(u) = \rho$ as claimed.   ∎

If f' were a continuous function then the theorem would just be a special instance of the Intermediate Value Property of continuous functions (see Corollary 6.18). But derivatives need not be continuous, as the example

$$f(x) = \begin{cases} x^2 \cdot \sin(1/x) & \text{if} \quad x \neq 0 \\ 0 & \text{if} \quad x = 0 \end{cases}$$

illustrates. Verify for yourself that $f'(0)$ exists and vanishes but $\lim_{x \to 0} f'(x)$ does not exist. This example illustrates the significance of the theorem. Since f' will always satisfy the Intermediate Value Property (even when it is not continuous), its discontinuities cannot be of the first kind. In other words:

> If f is a differentiable function on an open interval I then the discontinuities of f' are all of the second kind.

Next we turn to the simplest form of the Mean Value Theorem.

THEOREM 7.7   ROLLE'S THEOREM

Let f be a continuous function on the closed interval $[a, b]$, which is differentiable on (a, b). If $f(a) = f(b) = 0$ then there is a point $\xi \in (a, b)$ such that $f'(\xi) = 0$.

PROOF   If f is a constant function then any point ξ in the interval will do. So assume that f is nonconstant.

Theorem 6.14 guarantees that f will have both a maximum and a minimum in $[a, b]$. If one of these occurs in (a, b) then Proposition 7.5 guarantees that f' will vanish at that point and we are done. If both occur at the endpoints then all the values of f lie between 0 and 0. In other words, f is constant, contradicting our assumption. ∎

Of course the point ξ in Rolle's Theorem need not be unique. If $f(x) = x^3 - x^2 - 2x$ on the interval $[-1, 2]$ then $f(a) = f(b) = 0$ and f' vanishes at *two* points of the interval $(-1, 2)$.

If you rotate the graph of a function satisfying the hypotheses of Rolle's Theorem, the result suggests that for any continuous function f on an interval $[a, b]$, differentiable on (a, b), we should be able to relate the slope of the chord connecting $(a, f(a))$ and $(b, f(b))$ with the value of f' at some interior point. That is the content of the Mean Value Theorem:

THEOREM 7.8   THE MEAN VALUE THEOREM
Let f be a continuous function on the closed interval $[a, b]$ that is differentiable on (a, b). There exists a point $\xi \in (a, b)$ such that

$$\frac{f(b) - f(a)}{b - a} = f'(\xi).$$

PROOF   Our scheme is to implement the remarks preceding the theorem: we rotate the picture to reduce to the case of Rolle's Theorem. More precisely, define

$$g(x) = f(x) - \left[f(a) + \frac{f(b) - f(a)}{b - a} \cdot (x - a) \right] \quad \text{if} \quad x \in [a, b].$$

By direct verification, g is continuous on $[a, b]$ and differentiable on (a, b) (after all, g is obtained from f by elementary arithmetic operations). Also $g(a) = g(b) = 0$. Thus we may apply Rolle's Theorem to g, and we find that there is a $\xi \in (a, b)$ such that $g'(\xi) = 0$. Remembering that x is the variable, we differentiate the formula for g to find that

$$0 = g'(\xi) = \left[f'(x) - \frac{f(b) - f(a)}{b - a} \right]\Bigg|_{x=\xi}$$

$$= \left[f'(\xi) - \frac{f(b) - f(a)}{b - a} \right].$$

As a result,

$$f'(\xi) = \frac{f(b) - f(a)}{b - a}. \qquad ∎$$

COROLLARY 7.9

If f is a differentiable function on the open interval I and if $f'(x) = 0$ for all $x \in I$ then f is a constant function.

PROOF   If s and t are any two elements of I then the theorem tells us that

$$f(s) - f(t) = f'(\xi) \cdot (s - t)$$

for some ξ between s and t. By hypothesis, however, $f'(\xi) = 0$. We conclude that $f(s) = f(t)$. But since s and t were chosen arbitrarily we must conclude that f is constant.   ∎

COROLLARY 7.10

If f is differentiable on an open interval I and $f'(x) \geq 0$ for all $x \in I$ then f is **monotone increasing** *on I; that is, if $s < t$ are elements of I then $f(s) \leq f(t)$.*

If f is differentiable on an open interval I and $f'(x) \leq 0$ for all $x \in I$ then f is **monotone decreasing** *on I; that is, if $s < t$ are elements of I then $f(s) \geq f(t)$.*

PROOF   Similar to the preceding corollary.   ∎

Example 7.3

Let us verify that if f is a differentiable function on $\mathbb{R}$ and if $|f'(x)| \leq 1$ for all x then $|f(s) - f(t)| \leq |s - t|$ for all real s and t.

In fact, for $s \neq t$ there is a ξ between s and t such that

$$\frac{f(s) - f(t)}{s - t} = f'(\xi).$$

But $|f'(\xi)| \leq 1$ by hypothesis hence

$$\left| \frac{f(s) - f(t)}{s - t} \right| \leq 1$$

or

$$|f(s) - f(t)| \leq |s - t|.$$ □

Example 7.4

Let us verify that

$$\lim_{x \to +\infty} \left(\sqrt{x + 5} - \sqrt{x} \right) = 0.$$

Here the limit operation means that for any $\epsilon > 0$ there is an $N > 0$ such that $x > N$ implies that the expression in parentheses has absolute value less than ϵ.

Define $f(x) = \sqrt{x}$ for $x > 0$. Then the expression in parentheses is just $f(x+5) - f(x)$. By the Mean Value Theorem this equals

$$f'(\xi) \cdot 5$$

for some $x < \xi < x + 5$. But this last expression is

$$\frac{1}{2} \cdot \xi^{-1/2} \cdot 5.$$

By the bounds on ξ, this is

$$\leq \frac{5}{2} x^{-1/2}.$$

Clearly, as $x \to +\infty$, this expression tends to zero. $\quad \Box$

A powerful tool in analysis is a generalization of the usual Mean Value Theorem that is due to A. L. Cauchy:

THEOREM 7.11   CAUCHY'S MEAN VALUE THEOREM
Let f and g be continuous functions on the interval $[a, b]$ that are both differentiable on the interval (a, b). Then there is a point $\xi \in (a, b)$ such that

$$\frac{f(b) - f(a)}{g(b) - g(a)} = \frac{f'(\xi)}{g'(\xi)}.$$

PROOF   Apply the usual Mean Value Theorem to the function

$$h(x) = g(x) \cdot \{f(b) - f(a)\} - f(x) \cdot \{g(b) - g(a)\}. \quad \blacksquare$$

Clearly the usual Mean Value Theorem is obtained from Cauchy's by taking $g(x)$ to be the function x. We conclude this section by illustrating a typical application of the result.

Example 7.5
Let f be a differentiable function on an open interval I such that f' is differentiable at a point $x \in I$. Then

$$\lim_{h \to 0^+} \frac{2(f(x+h) + f(x-h) - 2f(x))}{h^2} = (f')'(x).$$

To see this, fix x and define $\mathcal{F}(h) = f(x+h)+f(x-h)-2f(x)$ and $\mathcal{G}(h) = h^2$. Then

$$\frac{2(f(x+h)+f(x-h)-2f(x))}{h^2} = \frac{\mathcal{F}(h)-\mathcal{F}(0)}{\mathcal{G}(h)-\mathcal{G}(0)}.$$

According to Cauchy's Mean Value Theorem, there is a ξ between 0 and h such that the last line equals

$$\frac{\mathcal{F}'(\xi)}{\mathcal{G}'(\xi)}.$$

Writing this expression out gives

$$\frac{f'(x+\xi)-f'(x-\xi)}{2\xi} = \frac{1}{2}\cdot\frac{f'(x+\xi)-f'(x)}{\xi}$$
$$+\frac{1}{2}\cdot\frac{f'(x-\xi)-f'(x)}{-\xi},$$

and as $h \to 0$ the last line tends, by the definition of the derivative, to the quantity $(f')'(x)$. $\square$

7.3 More on the Theory of Differentiation

l'Hôpital's Rule (actually due to his teacher J. Bernoulli (1667–1748)) is a useful device for calculating limits and a nice application of the Cauchy Mean Value Theorem. Here we present a special case of the theorem.

THEOREM 7.12
Suppose that f and g are differentiable functions on an open interval I and that $p \in I$. If $\lim_{x \to p} f(x) = \lim_{x \to p} g(x) = 0$ and if

$$\lim_{x \to p} \frac{f'(x)}{g'(x)} \tag{$*$}$$

exists and equals a real number ℓ then

$$\lim_{x \to p} \frac{f(x)}{g(x)} = \ell.$$

PROOF   Fix a real number $a > \ell$. By $(*)$ there is a number $q > p$ such that if $p < x < q$ then

$$\frac{f'(x)}{g'(x)} < a. \tag{$**$}$$

But now if $p < s < t < q$ then, by Cauchy's Mean Value Theorem,

$$\frac{f(t) - f(s)}{g(t) - g(s)} = \frac{f'(x)}{g'(x)}$$

for some $s < x < t$. It follows then from $(**)$ that

$$\frac{f(t) - f(s)}{g(t) - g(s)} < a.$$

Now let $s \to p$ and invoke the hypothesis about the zero limit of f and g at p to conclude that

$$\frac{f(t)}{g(t)} \leq a$$

when $p < t < q$. Since a is an arbitrary number to the right of ℓ we conclude that

$$\limsup_{t \to p^+} \frac{f(t)}{g(t)} \leq \ell.$$

Similar arguments show that

$$\liminf_{t \to p^+} \frac{f(t)}{g(t)} \geq \ell;$$

$$\limsup_{t \to p^-} \frac{f(t)}{g(t)} \leq \ell;$$

$$\liminf_{t \to p^-} \frac{f(t)}{g(t)} \geq \ell.$$

We conclude that the desired limit exists and equals ℓ. ∎

PROPOSITION 7.13
Let f be an invertible function on an interval (a, b) with nonzero derivative at a point $x \in (a, b)$. Let $X = f(x)$. Then $\left(f^{-1}\right)'(X)$ exists and equals $1/f'(x)$.

PROOF   Observe that, for $T \neq X$,

$$\frac{f^{-1}(T) - f^{-1}(X)}{T - X} = \frac{1}{\frac{f(t) - f(x)}{t - x}}, \qquad (*)$$

where $t = f^{-1}(T)$. Since $f'(x) \neq 0$, the difference quotients for f in the denominator are bounded from zero, hence the limit of the formula in $(*)$ exists. This proves that f^{-1} is differentiable at X and that the derivative equals $1/f'(x)$. ∎

Example 7.6

We know that the function $f(x) = x^k$, k a positive integer, is one-to-one and differentiable on the interval $(0, 1)$. Moreover, the derivative $k \cdot x^{k-1}$ never vanishes on that interval. Therefore the proposition applies and we find for $X \in (0, 1) = f((0, 1))$ that

$$\left(f^{-1} \right)' (X) = \frac{1}{f'(x)} = \frac{1}{f'(X^{1/k})}$$

$$= \frac{1}{k \cdot X^{1-1/k}} = \frac{1}{k} \cdot X^{\frac{1}{k}-1}.$$

In other words,

$$\left(X^{1/k} \right)' = \frac{1}{k} X^{\frac{1}{k}-1}.$$

         □

We conclude this section by saying a few words about higher derivatives. If f is a differentiable function on an open interval I then we may ask whether the function f' is differentiable. If it is, we denote its derivative by

$$f'' \quad \text{or} \quad f^{(2)} \quad \text{or} \quad \frac{d^2}{dx^2} f \quad \text{or} \quad \frac{d^2 f}{dx^2},$$

and call it the second derivative of f. Likewise, the derivative of the $(k-1)^{\text{th}}$ derivative, if it exists, is called the kth derivative and is denoted

$$f^{'' \cdots '} \quad \text{or} \quad f^{(k)} \quad \text{or} \quad \frac{d^k}{dx^k} f \quad \text{or} \quad \frac{d^k f}{dx^k}.$$

Observe that we cannot even consider whether $f^{(k)}$ exists at a point unless $f^{(k-1)}$ exists in a *neighborhood* of that point.

If f is k times differentiable on an open interval I and if each of the derivatives $f^{(1)}, f^{(2)}, \ldots, f^{(k)}$ is continuous on I then we call f k *times continuously differentiable* on I. Obviously there is some redundancy in this definition since the continuity of $f^{(j-1)}$ follows from the existence of $f^{(j)}$. Thus only the continuity of the last derivative $f^{(k)}$ need be checked. Continuously differentiable functions are useful tools in analysis. We denote the class of k times continuously differentiable functions on I by $C^k(I)$.

For $k = 1, 2, \ldots$ the function

$$f_k(x) = \begin{cases} x^{k+1} & \text{if } x \geq 0 \\ -x^{k+1} & \text{if } x < 0 \end{cases}$$

will be k times continuously differentiable on $\mathbb{R}$ but will fail to be $k+1$ times differentiable at $x = 0$. More dramatically, an analysis similar to the one we used on the Weierstrass nowhere-differentiable function shows that the function

$$g_k(x) = \sum_{j=1}^{\infty} \frac{3^j}{4^{j+jk}} \sin(4^j x)$$

is k times continuously differentiable on $\mathbb{R}$ but will not be $k+1$ times differentiable at any point (this function, with $k = 0$, was Weierstrass's original example).

A more refined notion of smoothness of functions is that of Lipschitz or Hölder continuity. If f is a function on an open interval I and $0 < \alpha \leq 1$ then we say that f satisfies a *Lipschitz condition* of order α on I if there is a constant M such that for all $s, t \in I$ we have

$$|f(s) - f(t)| \leq M \cdot |s - t|^\alpha.$$

Such a function is said to be of class $\text{Lip}_\alpha(I)$. Clearly a function of class Lip_α is uniformly continuous on I. For if $\epsilon > 0$ then we may take $\delta = (\epsilon/M)^{1/\alpha}$: then for $|s - t| < \alpha$ we have

$$|f(s) - f(t)| \leq M \cdot |s - t|^\alpha < M \cdot \epsilon/M = \epsilon.$$

Interestingly, when $\alpha > 1$ the class Lip_α contains only constant functions. For in this instance the inequality

$$|f(s) - f(t)| \leq M \cdot |s - t|^\alpha$$

leads to

$$\left| \frac{f(s) - f(t)}{s - t} \right| \leq M \cdot |s - t|^{\alpha - 1}.$$

Because $\alpha - 1 > 0$, letting $s \to t$ yields that $f'(t)$ exists for every $t \in I$ and equals 0. It follows from Corollary 7.9 of the last section that f is constant on I.

Instead of trying to extend the definition of $\text{Lip}_\alpha(I)$ to $\alpha > 1$ it is customary to define classes of functions $C^{k,\alpha}$, for $k = 0, 1, \ldots$ and $0 < \alpha \leq 1$, by the condition that f be of class C^k on I and that $f^{(k)}$ be an element of $\text{Lip}_\alpha(I)$. We leave it as an exercise for you to verify that $C^{k,\alpha} \subseteq C^{\ell,\beta}$ if either $k > \ell$ or both $k = \ell$ and $\alpha \geq \beta$.

In more advanced studies in analysis, it is appropriate to replace $\text{Lip}_1(I)$, and more generally $C^{k,1}$, with another space (the space of "smooth functions" invented by Antoni Zygmund, 1900–) defined in a more subtle fashion. These matters exceed the scope of this book, but we shall make a few remarks about them in the Exercises.

Exercises

7.1 Prove part (a) of Theorem 7.2.

7.2 If f is a C^2 function on $\mathbb{R}$ and if $|f''(x)| \leq C$ for all x then prove that

$$\left| \frac{f(x+h) + f(x-h) - 2f(x)}{h^2} \right| \leq C.$$

7.3 Give an example of a function f for which the limit in Example 7.5 exists at some x, but for which f is not twice differentiable at x.

7.4 For which positive integers k is it true that if $f^k = f \cdot f \cdots f$ is differentiable at x then f is differentiable at x?

7.5 In which class $C^{k,\alpha}$ is the function $x \cdot \ln|x|$ on the interval $[-1/2, 1/2]$? How about the function $x/\ln|x|$?

7.6 Give an example of a function on $\mathbb{R}$ such that

$$\left| \frac{f(x+h) + f(x-h) - 2f(x)}{h} \right| \leq C$$

for all x and all $h \neq 0$ but f is not in $\text{Lip}_1(\mathbb{R})$. (Hint: See Exercise 5.)

7.7 Fix a positive integer k. Give examples of two functions f and g, neither of which is in C^k but such that $f \cdot g \in C^k$.

7.8 Fix a positive integer ℓ and define $f(x) = |x|^\ell$ on the interval $(-1, 1)$. In which class C^k does f lie? In which class $C^{k,\alpha}$ does it lie?

7.9 Let f be a function that has domain an interval I and takes values in the complex numbers. Then we may write $f(x) = u(x) + iv(x)$ with u and v each being real-valued functions. We say that f is differentiable at a point $x \in I$ if both u and v are. Formulate an alternative definition of differentiability of f at a point x that makes no reference to u and v (but instead defines the derivative directly in terms of f), and prove that your new definition is equivalent to the definition in terms of u and v.

7.10 Refer to Exercise 9 for terminology. Verify the properties of the derivative presented in Theorem 7.2 in the new context of complex-valued functions.

7.11 Let f be a function that is continuous on $[0, \infty)$ and differentiable on $(0, \infty)$. If $f(0) = 0$ and $|f'(x)| \leq |f(x)|$ for all $x > 0$ then prove that $f(x) = 0$ for all x. [This result is often called Gronwall's inequality.]

7.12 Let $E \subseteq \mathbb{R}$ be a closed set. Fix a nonnegative integer k. Show that there is a function f in $C^k(\mathbb{R})$ such that $E = \{x : f(x) = 0\}$.

7.13 Prove that the nowhere-differentiable function constructed in Theorem 7.3 is in Lip_α for all $\alpha < 1$.

7.14 Let f be a continuous function on $[a, b]$ that is differentiable on (a, b). Assume that $f(a) = m$ and that $|f'(x)| \leq K$ for all $x \in (a, b)$. What bound can you then put on the magnitude of $f(b)$?

7.15 Let f be a differentiable function on an open interval I and assume that f has no local minima nor local maxima on I. Prove that f is either monotone increasing or monotone decreasing on I.

7.16 Let f be a differentiable function on an open interval I. Prove that f' is continuous if and only if the inverse image under f' of any point is a closed set.

7.17 Let $f(x)$ equal 0 if x is irrational; let $f(x)$ equal $1/q$ if x is a rational number that can be expressed in lowest terms as p/q. Is f differentiable at any x?

7.18 In the text we give sufficient conditions for the inclusion $C^{k,\alpha} \subseteq C^{\ell,\beta}$. Show that the inclusion is strict if either $k > l$ or $k = \ell$ and $\alpha > \beta$.

7.19 If $0 < \alpha \leq 1$ then prove that there is a constant $C_\alpha > 0$ such that for $0 < x < 1/2$ it holds that

$$|\ln x| \leq C_\alpha \cdot x^{-\alpha}.$$

Prove that the constant cannot be taken to be independent of α.

7.20 If a function f is twice differentiable on $(0, \infty)$ and $f''(x) \geq c > 0$ for all x then prove that f is not bounded from above.

7.21 If f is differentiable on an interval I and $f'(x) > 0$ for all $x \in I$ then does it follow that $(f^2)' > 0$ for all $x \in I$? What additional hypothesis on f will make the conclusion true?

7.22 Answer Exercise 21 with the exponent 2 replaced by any positive integer exponent.

7.23 Suppose that f is a differentiable function on an interval I and that $f'(x)$ is never zero. Prove that f is invertible. Then prove that f^{-1} is differentiable. Finally, use the Chain Rule on the identity $f(f^{-1}) = x$ to derive a formula for $(f^{-1})'$.

7.24 Assume that f is a continuous function on $(-1, 1)$ and that f is differentiable on $(-1, 0) \cup (0, 1)$. If the limit $\lim_{x \to 0} f'(x)$ exists, then is f differentiable at $x = 0$?

7.25 Formulate notions of "left differentiable" and "right differentiable" for functions defined on suitable half-open intervals. Also formulate definitions of "left continuous" and "right continuous." If you have done things correctly, then you should be able to prove that a left-differentiable (resp. right-differentiable) function is left continuous (resp. right continuous).

8

The Integral

8.1 Partitions and the Concept of Integral

We learn in calculus that it is often useful to think of an integral as representing area. However, this is but one of many important applications of integration theory. The integral is a generalization of the summation process. That is the point of view that we shall take in this chapter.

DEFINITION 8.1 *Let $[a, b]$ be a closed interval in $\mathbb{R}$. A finite, ordered set of points $\mathcal{P} = \{x_0, x_1, x_2, \ldots, x_{k-1}, x_k\}$ such that*

$$a = x_0 \leq x_1 \leq x_2 \leq \ldots \leq x_{k-1} \leq x_k = b$$

*is called a **partition** of $[a, b]$. Refer to the figure.*

*If $\mathcal{P}$ is a partition of $[a, b]$ then we let I_j denote the interval $[x_{j-1}, x_j]$, $j = 1, 2, \ldots, k$. The symbol Δ_j denotes the **length** of I_j. The **mesh** of $\mathcal{P}$, denoted by $m(\mathcal{P})$, is defined to be $\max \Delta_j$.*

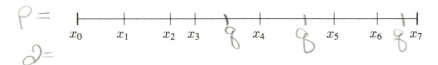

The points of a partition need not be equally spaced, nor must they be distinct from each other.

Example 8.2
The set $\mathcal{P} = \{0, 1, 1, 9/8, 2, 5, 21/4, 23/4, 6\}$ is a partition of the interval $[0, 6]$ with mesh 3 (because $I_5 = [2, 5]$, with length 3, is the longest interval in the partition). □

DEFINITION 8.3 *Let* $[a, b]$ *be an interval and let* f *be a function with domain* $[a, b]$. *If* $\mathcal{P} = \{x_0, x_1, x_2, \ldots, x_{k-1}, x_k\}$ *is a partition of* $[a, b]$ *and if, for each* j, s_j *is an element of* I_j *then the corresponding* **Riemann sum** *is defined to be*

$$\mathcal{R}(f, \mathcal{P}) = \sum_{j=1}^{k} f(s_j) \Delta_j.$$

Example 8.4

Let $f(x) = x^2 - x$ and $[a, b] = [1, 4]$. Define the partition $\mathcal{P} = \{1, 3/2, 2, 7/3, 4\}$ of this interval. Then a Riemann sum for this f and $\mathcal{P}$ is

$$\mathcal{R}(f, \mathcal{P}) = \left(1^2 - 1\right) \cdot \frac{1}{2} + \left((7/4)^2 - (7/4)\right) \cdot \frac{1}{2}$$
$$+ \left((7/3)^2 - (7/3)\right) \cdot \frac{1}{3} + \left(3^2 - 3\right) \cdot \frac{5}{3}$$
$$= \frac{10103}{864}.$$

Notice that we have complete latitude in choosing each point s_j from the corresponding interval I_j. While at first confusing, we will find this freedom to be a powerful tool when proving results about the integral.

The first main step in the theory of the Riemann integral is to determine a method for "calculating the limit of the Riemann sums" of a function as the mesh of partitions tends to zero. There are in fact several methods for doing this. We have chosen the simplest one.

DEFINITION 8.5 *Let* $[a, b]$ *be an interval and* f *a function with domain* $[a, b]$. *We say that* **the Riemann sums of** f **tend to a limit** ℓ **as** $m(\mathcal{P})$ **tends to** 0 *if for any* $\epsilon > 0$ *there is a* $\delta > 0$ *such that if* $\mathcal{P}$ *is any partition of* $[a, b]$ *with* $m(\mathcal{P}) < \delta$ *then* $|\mathcal{R}(f, \mathcal{P}) - \ell| < \epsilon$ *for every choice of* $s_j \in I_j$.

It will turn out to be critical for the success of this definition that we require that *every* partition of mesh smaller than δ satisfy the conclusion of the definition. The theory does not work if for every $\epsilon > 0$ there is a $\delta > 0$ and some partition $\mathcal{P}$ of mesh less than δ that satisfies the conclusion of the definition.

DEFINITION 8.6 *A function* f *on a closed interval* $[a, b]$ *is said to be* **Riemann integrable** *on* $[a, b]$ *if the Riemann sums of* $\mathcal{R}(f, \mathcal{P})$ *tend to a limit as* $m(\mathcal{P})$ *tends to zero.*

$$\mathcal{L}(\varphi) = \sum f(x_1)[x_2 - x_1]$$
$$u(f,P) = \sum f(x_2)[x_2 - x_1]$$

*The value of the limit, when it exists, is called the **Riemann integral** of f over $[a, b]$ and is denoted by*

$$\int_a^b f(x)\, dx.$$

REMARK 8.1 We mention now a useful fact that will be formalized in later sections. Suppose that f is Riemann integrable on $[a, b]$ with the value of the integral being ℓ. Let $\epsilon > 0$. Then, as stated in the definition (with $\epsilon/2$ replacing ϵ), there is a $\delta > 0$ such that if $\mathcal{Q}$ is a partition of $[a, b]$ of mesh smaller than δ then $|\mathcal{R}(f, \mathcal{Q}) - \ell| < \epsilon/2$. It follows that if $\mathcal{P}$ and $\mathcal{P}'$ are partitions of $[a, b]$ of mesh smaller than δ then

$$|\mathcal{R}(f, \mathcal{P}) - \mathcal{R}(f, \mathcal{P}')| \le |\mathcal{R}(f, \mathcal{P}) - \ell| + |\ell - \mathcal{R}(f, \mathcal{P}')| < \frac{\epsilon}{2} + \frac{\epsilon}{2} = \epsilon.$$

Note, however, that we may choose $\mathcal{P}'$ to equal the partition $\mathcal{P}$. Also we may for each j choose the points s_j, where f is evaluated for the Riemann sum over $\mathcal{P}$, to be a point where f very nearly assumes its supremum on I_j. Then we may for each j choose the points s_j', where f is evaluated for the Riemann sum over $\mathcal{P}'$, to be a point where f very nearly assumes its infimum on I_j. It easily follows that when the mesh of $\mathcal{P}$ is less than δ then

$$\sum_j \left(\sup_{I_j} f - \inf_{I_j} f \right) \Delta_j \le \epsilon. \qquad (*)$$

This consequence of integrability will prove useful to us in some of the discussions in this and the next section. In the Exercises we shall consider in detail the assertion that integrability implies $(*)$ and the converse as well. ∎

DEFINITION 8.7 *If $\mathcal{P}, \mathcal{P}'$ are partitions of $[a, b]$, then their **common refinement** is the union of all the points of $\mathcal{P}$ and $\mathcal{P}'$.*

We record now a technical lemma that will be used in several of the proofs that follow:

LEMMA 8.2
Let f be a function with domain the closed interval $[a, b]$. The Riemann integral

$$\int_a^b f(x)\, dx$$

exists if and only if for every $\epsilon > 0$ there is a $\delta > 0$ such that if $\mathcal{P}$ and $\mathcal{P}'$ are partitions of $[a, b]$ with $m(\mathcal{P}) < \delta$ and $m(\mathcal{P}') < \delta$ then their common refinement

Q has the property that

$$|\mathcal{R}(f, \mathcal{P}) - \mathcal{R}(f, \mathcal{Q})| < \epsilon$$

and $\qquad\qquad\qquad\qquad\qquad\qquad\qquad\qquad\qquad\qquad\qquad\qquad (*)$

$$|\mathcal{R}(f, \mathcal{P}') - \mathcal{R}(f, \mathcal{Q})| < \epsilon.$$

PROOF If f is Riemann integrable then the assertion of the lemma follows immediately from the definition of the integral.

For the converse, note that $(*)$ certainly implies that if $\epsilon > 0$ then there is a $\delta > 0$ such that if $\mathcal{P}$ and $\mathcal{P}'$ are partitions of $[a, b]$ with $m(\mathcal{P}) < \delta$ and $m(\mathcal{P}') < \delta$ then

$$|\mathcal{R}(f, \mathcal{P}) - \mathcal{R}(f, \mathcal{P}')| < \epsilon \qquad\qquad\qquad (**)$$

(just use the triangle inequality).

Now for each $\epsilon_j = 2^{-j}, j = 1, 2, \ldots$, we can choose a $\delta_j > 0$ as in $(**)$. Set S_j be the *closure* of the set

$$\{\mathcal{R}(f, \mathcal{P}) : m(\mathcal{P}) < \delta_j\}.$$

By the choice of δ_j, S_j is contained in an interval of length not greater than $2\epsilon_j$. On the one hand,

$$\bigcap_j S_j$$

must be nonempty since it is the decreasing intersection of compact sets. On the other hand, the estimate implies that the intersection must be contained in a closed interval of length 0 — that is, the intersection is a point. That point is then the limit of the Riemann sums, that is, the value of the Riemann integral. ∎

The most important, and perhaps the simplest, fact about the Riemann integral is that a large class of familiar functions is Riemann integrable:

THEOREM 8.3
Let f be a continuous function on a nonempty closed interval $[a, b]$. Then f is Riemann integrable on $[a, b]$.

PROOF We use the lemma. Assume that $a < b$. Given $\epsilon > 0$, choose (by the uniform continuity of f on I — Theorem 6.15) a $\delta > 0$ such that whenever $|s - t| < \delta$ then

$$|f(s) - f(t)| < \frac{\epsilon}{b - a}. \qquad\qquad\qquad (*)$$

Let $\mathcal{P}$ and $\mathcal{P}'$ be any two partitions of $[a, b]$ of mesh smaller than δ. Let $\mathcal{Q}$ be the common refinement of $\mathcal{P}$ and $\mathcal{P}'$.

Now we let I_j denote the intervals arising in the partition $\mathcal{P}$ (and having length Δ_j) and $\tilde{I}_\ell$ the intervals arising in the partition $\mathcal{Q}$ (and having length $\tilde{\Delta}_\ell$). Since the partition $\mathcal{Q}$ contains every point of $\mathcal{P}$, plus some additional points as well, every $\tilde{I}_\ell$ is contained in some I_j. Fix j and consider the expression

$$\left| f(s_j)\Delta_j - \sum_{\tilde{I}_\ell \subseteq I_j} f(t_\ell)\tilde{\Delta}_\ell \right|. \qquad (\ast\ast)$$

We write

$$\Delta_j = \sum_{\tilde{I}_\ell \subseteq I_j} \tilde{\Delta}_\ell.$$

This equality enables us to rearrange $(\ast\ast)$ as

$$\left| f(s_j) \cdot \sum_{\tilde{I}_\ell \subseteq I_j} \tilde{\Delta}_\ell - \sum_{\tilde{I}_\ell \subseteq I_j} f(t_\ell)\tilde{\Delta}_\ell \right| = \left| \sum_{\tilde{I}_\ell \subseteq I_j} [f(s_j) - f(t_\ell)]\tilde{\Delta}_\ell \right|$$

$$\leq \sum_{\tilde{I}_\ell \subseteq I_j} |f(s_j) - f(t_\ell)|\tilde{\Delta}_\ell.$$

But each of the points t_ℓ is in the interval I_j, as is s_j. So they differ by less than δ. Therefore, by $(\ast)$, the last expression is less than

$$\sum_{\tilde{I}_\ell \subseteq I_j} \frac{\epsilon}{b-a}\tilde{\Delta}_\ell = \frac{\epsilon}{b-a} \sum_{\tilde{I}_\ell \subseteq I_j} \tilde{\Delta}_\ell = \frac{\epsilon}{b-a} \cdot \Delta_j.$$

Now we conclude the argument by writing

$$|\mathcal{R}(f,\mathcal{P}) - \mathcal{R}(f,\mathcal{Q})| = \left| \sum_j f(s_j)\Delta_j - \sum_\ell f(t_\ell)\tilde{\Delta}_\ell \right|$$

$$\leq \sum_j \left| f(s_j)\Delta_j - \sum_{\tilde{I}_\ell \subseteq I_j} f(t_\ell)\tilde{\Delta}_\ell \right|$$

$$< \sum_j \frac{\epsilon}{b-a} \cdot \Delta_j$$

$$= \frac{\epsilon}{b-a} \cdot \sum_j \Delta_j$$

$$= \frac{\epsilon}{b-a} \cdot (b-a)$$

$$= \epsilon.$$

The estimate for $|\mathcal{R}(f,\mathcal{P}') - \mathcal{R}(f,\mathcal{Q})|$ is identical and we omit it. The result now follows from Lemma 8.2. ∎

In the exercises we will ask you to extend the theorem to the case of functions f on $[a, b]$ that are bounded and have finitely many, or even countably many, discontinuities.

We conclude this section by noting an important fact about Riemann integrable functions. A Riemann integrable function on an interval $[a, b]$ *must be bounded.* If it were not, then one could choose the points s_j in the construction of $\mathcal{R}(f, \mathcal{P})$ so that $f(s_j)$ is arbitrarily large, and the Riemann sums would become arbitrarily large, hence could not converge. You will be asked in the Exercises to work out the details of this assertion.

8.2 Properties of the Riemann Integral

We begin this section with a few elementary properties of the integral that reflect its linear nature.

THEOREM 8.4

Let $[a, b]$ be a nonempty interval, let f and g be Riemann integrable functions on the interval, and let α be a real number. Then $f \pm g$ and $\alpha \cdot f$ are integrable and we have

(a) $\int_a^b f(x) \pm g(x)\, dx = \int_a^b f(x)\, dx \pm \int_a^b g(x)\, dx$;

(b) $\int_a^b \alpha \cdot f(x)\, dx = \alpha \cdot \int_a^b f(x)\, dx.$

PROOF For (a), let

$$A = \int_a^b f(x)\, dx$$

and

$$B = \int_a^b g(x)\, dx.$$

Let $\epsilon > 0$. Choose a $\delta_1 > 0$ such that if $\mathcal{P}$ is a partition of $[a, b]$ with mesh less than δ_1 then

$$|\mathcal{R}(f, \mathcal{P}) - A| < \frac{\epsilon}{2}.$$

Similarly, choose a $\delta_2 > 0$ such that if $\mathcal{P}$ is a partition of $[a, b]$ with mesh less than δ_2 then

$$|\mathcal{R}(f, \mathcal{P}) - B| < \frac{\epsilon}{2}.$$

Let $\delta = \min\{\delta_1, \delta_2\}$. If $\mathcal{P}'$ is any partition of $[a, b]$ with $m(\mathcal{P}') < \delta$ then

$$
\begin{aligned}
|\mathcal{R}(f+g, \mathcal{P}') - (A+B)| &= |\mathcal{R}(f, \mathcal{P}') + \mathcal{R}(g, \mathcal{P}') - (A+B)| \\
&\leq |\mathcal{R}(f, \mathcal{P}') - A| + |\mathcal{R}(g, \mathcal{P}') - B| \\
&< \frac{\epsilon}{2} + \frac{\epsilon}{2} = \epsilon
\end{aligned}
$$

This means that the integral of $f + g$ exists and equals $A + B$, as we were required to prove.

The proof of (b) follows similar lines but is much easier and we leave it as an exercise. ∎

THEOREM 8.5

If c is a point of the interval $[a, b]$ and if f is Riemann integrable on both $[a, c]$ and $[c, b]$ then f is integrable on $[a, b]$ and $\int_a^c f(x)\, dx + \int_b^c f(x)\, dx = \int_a^b f(x)\, dx$.

PROOF   Let us write

$$
A = \int_a^c f(x)\, dx
$$

and

$$
B = \int_c^b f(x)\, dx.
$$

Now pick $\epsilon > 0$. There is a $\delta_1 > 0$ such that if $\mathcal{P}$ is a partition of $[a, c]$ with mesh less than δ_1 then

$$
|\mathcal{R}(f, \mathcal{P}) - A| < \frac{\epsilon}{3}.
$$

Similarly, choose $\delta_2 > 0$ such that if $\mathcal{P}'$ is a partition of $[c, b]$ with mesh less than δ_2 then

$$
|\mathcal{R}(f, \mathcal{P}') - B| < \frac{\epsilon}{3}.
$$

Let M be an upper bound for $|f|$ (recall, from the last paragraph of Section 8.1, that a Riemann integrable function must be bounded). Set $\delta = \min\{\delta_1, \delta_2, \epsilon/(6M)\}$. Now let $\mathcal{V} = \{v_1, \ldots, v_k\}$ be any partition of $[a, b]$ with mesh less than δ. There is a last point v_n that is in $[a, c]$ and a first point v_{n+1} in $[c, b]$. Observe that $\mathcal{P} = \{v_0, \ldots, v_n, c\}$ is a partition of $[a, c]$ with mesh smaller than δ_1 and $\mathcal{P}' = \{c, v_{n+1}, \ldots, v_k\}$ is a partition of $[c, b]$ with mesh smaller than δ_2. Notice that $p_{n+1} = p_0' = c$. For each j let s_j be a point chosen in the interval $I_j = [v_{j-1}, v_j]$ from the partition $\mathcal{V}$. Then we have

$$
\begin{aligned}
|\mathcal{R}(f, \mathcal{V}) - [A + B]| &= \left| \left(\sum_{j=1}^{n} f(s_j)\Delta_j - A \right) \right. \\
&\quad + f(s_{n+1})\Delta_{n+1} + \left. \left(\sum_{j=n+2}^{k} f(s_j)\Delta_j - B \right) \right| \\
&= \left| \left(\sum_{j=1}^{n} f(s_j)\Delta_j + f(c) \cdot (c - v_n) - A \right) \right. \\
&\quad + \left(f(c) \cdot (v_{n+1} - c) + \sum_{j=n+2}^{k} f(s_j)\Delta_j - B \right) \\
&\quad + \big(f(s_{n+1}) - f(c) \big) \cdot (c - v_n) \\
&\quad + \left. \big(f(s_{n+1}) - f(c) \big) \cdot (v_{n+1} - c) \right| \\
&\leq \left| \left(\sum_{j=1}^{n} f(s_j)\Delta_j + f(c) \cdot (c - v_n) - A \right) \right| \\
&\quad + \left| \left(f(c) \cdot (v_{n+1} - c) + \sum_{j=n+2}^{k} f(s_j)\Delta_j - B \right) \right| \\
&\quad + |(f(s_{n+1}) - f(c)) \cdot (v_{n+1} - v_n)| \\
&= |\mathcal{R}(f, \mathcal{P}) - A| + |\mathcal{R}(f, \mathcal{P}') - B| \\
&\quad + |(f(s_{n+1}) - f(c)) \cdot (v_{n+1} - v_n)| \\
&< \frac{\epsilon}{3} + \frac{\epsilon}{3} + 2M \cdot \delta \\
&\leq \epsilon
\end{aligned}
$$

by the choice of δ.

This shows that f is integrable on the entire interval $[a, b]$ and the value of the integral is

$$
A + B = \int_a^c f(x)\, dx + \int_c^b f(x)\, dx. \qquad \blacksquare
$$

REMARK 8.6   If we adopt the convention that

$$\int_b^a f(x)\, dx = -\int_a^b f(x)\, dx$$

(which is consistent with the way that the integral was defined in the first place), then Theorem 8.5 is true even when c is not an element of $[a, b]$. For instance, suppose that $c < a < b$. Then, by Theorem 8.5,

$$\int_c^a f(x)\, dx + \int_a^b f(x)\, dx = \int_c^b f(x)\, dx.$$

But this may be rearranged to read

$$\int_a^b f(x)\, dx = -\int_c^a f(x)\, dx + \int_c^b f(x)\, dx = \int_a^c f(x)\, dx + \int_c^b f(x)\, dx.\ \blacksquare$$

One of the basic tools of analysis is to perform estimates. Thus we require certain fundamental inequalities about integrals. These are recorded in the next theorem.

THEOREM 8.7
Let f and g be integrable functions on a nonempty interval $[a, b]$. Then

(i) $\left|\int_a^b f(x)\, dx\right| \leq \int_a^b |f(x)|\, dx;$

(ii) *If $f(x) \leq g(x)$ for all $x \in [a, b]$ then $\int_a^b f(x)\, dx \leq \int_a^b g(x)\, dx$.*

PROOF   If $\mathcal{P}$ is any partition of $[a, b]$ then

$$|\mathcal{R}(f, \mathcal{P})| \leq \mathcal{R}(|f|, \mathcal{P}).$$

The first assertion follows.
     Next,

$$\mathcal{R}(f, \mathcal{P}) \leq \mathcal{R}(g, \mathcal{P}).$$

This inequality implies the second assertion. $\blacksquare$

Another fundamental operation in the theory of the integral is "change of variable" (sometimes called the "u-substitution" in calculus books). We next turn to a careful formulation and proof of this operation. First we need a lemma:

LEMMA 8.8
If f is a Riemann integrable function on $[a, b]$ and if ϕ is a continuous function on a compact interval that contains the range of f then $\phi \circ f$ is Riemann integrable.

PROOF Let $\epsilon > 0$. Since ϕ is a continuous function on a compact set, it is uniformly continuous (Theorem 6.15). Let $\delta > 0$ be selected such that (i) $\delta < \epsilon$ and (ii) if $|x - y| < \delta$ then $|\phi(x) - \phi(y)| < \epsilon$.

Now the hypothesis that f is Riemann integrable implies that there exists a $\tilde{\delta} > 0$ such that if $\mathcal{P}$ and $\mathcal{P}'$ are partitions of $[a, b]$ and $m(\mathcal{P}), m(\mathcal{P}') < \tilde{\delta}$, then for the common refinement $\mathcal{Q}$ of $\mathcal{P}$ and $\mathcal{P}'$ it holds that

$$|\mathcal{R}(f, \mathcal{P}) - \mathcal{R}(f, \mathcal{Q})| < \delta^2.$$

Fix such a $\mathcal{P}, \mathcal{P}'$ and $\mathcal{Q}$. Let J_ℓ be the intervals of $\mathcal{Q}$ and I_j the intervals of $\mathcal{P}$. Each J_ℓ is contained in some $I_{j(\ell)}$. We write

$$|\mathcal{R}(\phi \circ f, \mathcal{P}) - \mathcal{R}(\phi \circ f, \mathcal{Q})| = \left| \sum_j \phi \circ f(t_j) \Delta_j - \sum_\ell \phi \circ f(s_\ell) \Delta_\ell \right|$$

$$= \left| \sum_j \sum_{J_\ell \subseteq I_j} \phi \circ f(t_j) \Delta_\ell - \sum_j \sum_{J_\ell \subseteq I_j} \phi \circ f(s_\ell) \Delta_\ell \right|$$

$$= \left| \sum_j \sum_{J_\ell \subseteq I_j} \left(\phi \circ f(t_j) - \phi \circ f(s_\ell) \right) \Delta_\ell \right|$$

$$\leq \left| \sum_j \sum_{J_\ell \subseteq I_j, \ell \in G} \left(\phi \circ f(t_j) - \phi \circ f(s_\ell) \right) \Delta_\ell \right|$$

$$+ \left| \sum_j \sum_{J_\ell \subseteq I_j, \ell \in B} \left(\phi \circ f(t_j) - \phi \circ f(s_\ell) \right) \Delta_\ell \right|$$

where we put ℓ in G if $J_\ell \subseteq I_{j(\ell)}$ and $0 \leq \left(\sup_{I_{j(\ell)}} f - \inf_{I_{j(\ell)}} f \right) < \delta$; otherwise we put ℓ into B. Notice that

$$\sum_{\ell \in B} \delta \Delta_\ell \leq \sum_{\ell \in B} \left(\sup_{I_{j(\ell)}} f - \inf_{I_{j(\ell)}} f \right) \cdot \Delta_\ell$$

$$\leq \sum_j \sum_{J_\ell \subseteq I_j} \left(\sup_{I_j} f - \inf_{I_j} f \right) \cdot \Delta_\ell$$

$$= \sum_j \left(\sup_{I_j} f - \inf_{I_j} f \right) \Delta_j$$

$$\leq \delta^2$$

by the choice of $\tilde{\delta}$ (and Remark 8.1). Therefore

$$\sum_{\ell \in B} \Delta_\ell \leq \delta.$$

Let M be an upper bound for $|\phi|$ (Corollary 6.13). Then

$$\left| \sum_j \sum_{J_\ell \subseteq I_j, \ell \in B} (\phi \circ f(t_j) - \phi \circ f(s_\ell)) \Delta_\ell \right| \leq \left| \sum_j \sum_{J_\ell \subseteq I_j, \ell \in B} (2 \cdot M) \Delta_\ell \right|$$

$$\leq 2 \cdot \delta \cdot M$$

$$< 2M\epsilon.$$

Also

$$\left| \sum_j \sum_{J_\ell \subseteq I_j, \ell \in G} (\phi \circ f(t_j) - \phi \circ f(s_\ell)) \Delta_\ell \right| \leq \left| \sum_j \sum_{J_\ell \subseteq I_j, \ell \in G} \epsilon \Delta_\ell \right|$$

since, for $\ell \in G$, we know that $|f(\alpha) - f(\beta)| < \delta$ for any $\alpha, \beta \in I_{j(\ell)}$. Therefore the last line does not exceed $(b - a) \cdot \epsilon$. Putting together our estimates, we find that

$$|\mathcal{R}(\phi \circ f, \mathcal{P}) - \mathcal{R}(\phi \circ f, \mathcal{Q})| < \epsilon \cdot (2M + (b - a)).$$

By symmetry, an analogous inequality holds for $\mathcal{P}'$. By Lemma 8.2, this is what we needed to prove.   ∎

An easier result is that if f is Riemann integrable on an interval $[a, b]$ and if $\mu : [\alpha, \beta] \to [a, b]$ is continuous, then $f \circ \mu$ is Riemann integrable. The proof of this assertion is assigned in the Exercises.

COROLLARY 8.9
If f and g are Riemann integrable on $[a, b]$, then so is the function $f \cdot g$.

PROOF   By Theorem 8.4, $f + g$ is integrable. By the lemma, $(f + g)^2 = f^2 + 2f \cdot g + g^2$ is integrable. But the lemma also implies that f^2 and g^2 are integrable (here we use the function $\phi(x) = x^2$). It results, by subtraction, that $2 \cdot f \cdot g$ is integrable. Hence $f \cdot g$ is integrable.   ∎

THEOREM 8.10
Let f be an integrable function on an interval $[a, b]$ of positive length. Let ψ be a continuously differentiable function from another interval $[\alpha, \beta]$ of positive length into $[a, b]$. Assume that ψ is monotone increasing, one-to-one, and onto.

Then

$$\int_a^b f(x)\,dx = \int_\alpha^\beta f(\psi(x)) \cdot \psi'(x)\,dx.$$

PROOF   We may assume that $\alpha < \beta$. Since f is integrable, its absolute value is bounded by some number M. Fix $\epsilon > 0$. Since ψ' is continuous on the compact interval $[\alpha, \beta]$, it is uniformly continuous (Theorem 6.15). Hence we may choose $\delta > 0$ so small that if $|s - t| < \delta$ then $|\psi'(s) - \psi'(t)| < \epsilon/\left(M \cdot (\beta - \alpha)\right)$. If $\mathcal{P} = \{p_0, \ldots, p_k\}$ is any partition of $[a, b]$ then there is an associated partition $\tilde{\mathcal{P}} = \{\psi^{-1}(p_0), \ldots, \psi^{-1}(p_k)\}$ of $[\alpha, \beta]$. For simplicity, denote the points of $\tilde{\mathcal{P}}$ by $\tilde{p}_j$. Let us choose the partition $\mathcal{P}$ so fine that the mesh of $\tilde{\mathcal{P}}$ is less than δ. If t_j are points of $I_j = [p_{j-1}, p_j]$ then there are corresponding points $s_j = \psi^{-1}(t_j)$ of $\tilde{I}_j = [\tilde{p}_{j-1}, \tilde{p}_j]$. Then we have

$$\sum_{j=1}^k f(t_j)\Delta_j = \sum_{j=1}^k f(t_j)(p_j - p_{j-1})$$

$$= \sum_{j=1}^k f(\psi(s_j))(\psi(\tilde{p}_j) - \psi(\tilde{p}_{j-1}))$$

$$= \sum_{j=1}^k f(\psi(s_j))\psi'(u_j)(\tilde{p}_j - \tilde{p}_{j-1}),$$

where we have used the Mean Value Theorem in the last line to find each u_j. Our problem at this point is that $f \circ \psi$ and ψ' are evaluated at different points. So we must do some estimation to correct that problem.

   The last displayed line equals

$$\sum_{j=1}^k f(\psi(s_j))\psi'(s_j)(\tilde{p}_j - \tilde{p}_{j-1}) + \sum_{j=1}^k f(\psi(s_j))\left(\psi'(u_j) - \psi'(s_j)\right)(\tilde{p}_j - \tilde{p}_{j-1}).$$

The first sum is a Riemann sum for $f(\psi(x)) \cdot \psi'(x)$ and the second sum is an error term. Since the points u_j and s_j are elements of the same interval $\tilde{I}_j$ of length less than δ, we conclude that $|\psi'(u_j) - \psi'(s_j)| < \epsilon/(M \cdot |\beta - \alpha|)$. Thus the error term in absolute value does not exceed

$$\sum_{j=1}^k M \cdot \frac{\epsilon}{M \cdot |\beta - \alpha|} \cdot (\tilde{p}_j - \tilde{p}_{j-1}) = \frac{\epsilon}{\beta - \alpha} \sum_{j=0}^k (\tilde{p}_j - \tilde{p}_{j-1}) = \epsilon.$$

This shows that every Riemann sum for f on $[a, b]$ with sufficiently small mesh corresponds to a Riemann sum for $f(\psi(x)) \cdot \psi'(x)$ on $[\alpha, \beta]$ plus an error term of size less than ϵ. A similar argument shows that every Riemann sum for $f(\psi(x)) \cdot \psi'(x)$ on $[\alpha, \beta]$ with sufficiently small mesh corresponds to a Riemann sum for f on $[a, b]$ plus an error term of magnitude less than ϵ. The conclusion is then that the integral of f on $[a, b]$ (which exists by hypothesis) and the integral of $f(\psi(x)) \cdot \psi'(x)$ on $[\alpha, \beta]$ (which exists by the corollary to the lemma) agree.

∎

We conclude this section with the very important

THEOREM 8.11   THE FUNDAMENTAL THEOREM OF CALCULUS
Let f be an integrable function on the interval $[a, b]$. For $x \in [a, b]$ we define

$$F(x) = \int_a^x f(s) \, ds.$$

If f is continuous at $x \in (a, b)$ then

$$F'(x) = f(x).$$

PROOF   Fix $x \in (a, b)$. Let $\epsilon > 0$. Choose, by the continuity of f at x, a $\delta > 0$ such that $|s - x| < \delta$ implies $|f(s) - f(x)| < \epsilon$. We may assume that $\delta < \min\{x - a, b - x\}$. If $0 < |t - x| < \delta$ then

$$\left| \frac{F(t) - F(x)}{t - x} - f(x) \right| = \left| \frac{\int_a^t f(s) \, ds - \int_a^x f(s) \, ds}{t - x} - f(x) \right|$$

$$= \left| \frac{\int_x^t f(s) \, ds}{t - x} - \frac{\int_x^t f(x) \, ds}{t - x} \right|$$

$$= \left| \frac{\int_x^t \left(f(s) - f(x) \right) \, dx}{t - x} \right|.$$

Notice that we rewrote $f(x)$ as the integral with respect to a dummy variable s over an interval of length $|t - x|$ divided by $(t - x)$. Assume for the moment that $t > x$. Then the last line is dominated by

$$\frac{\int_x^t |f(s) - f(x)| \, ds}{t - x} \leq \frac{\int_x^t \epsilon \, ds}{t - x} = \epsilon.$$

A similar estimate holds when $t < x$ (simply reverse the limits of integration). This shows that

$$\lim_{t \to x} \frac{F(t) - F(x)}{t - x}$$

exists and equals $f(x)$. Thus $F'(x)$ exists and equals $f(x)$.   ∎

In the Exercises we shall consider how to use the theory of one-sided limits to make the conclusion of the Fundamental Theorem true on the entire interval $[a, b]$. We conclude with

COROLLARY 8.12
If f is a continuous function on $[a, b]$ and if G is any continuous function on $[a, b]$, differentiable on (a, b), whose derivative equals f on (a, b) then

$$\int_a^b f(x)\, dx = G(b) - G(a).$$

PROOF Define F as in the theorem. Since F and G have the same derivative on (a, b), they differ by a constant (see Corollary 7.9). Then

$$\int_a^b f(x)\, dx = F(b) = F(b) - F(a) = G(b) - G(a)$$

as desired. ∎

8.3 Another Look at the Integral

For many purposes, such as integration by parts, it is natural to formulate the integral in a more general context than we have considered in the first two sections. Our new formulation is called the *Riemann–Stieltjes integral* and is described below.

Fix an interval $[a, b]$ and a monotonically increasing function α on $[a, b]$. If $\mathcal{P} = \{p_0, p_1, \ldots, p_k\}$ is a partition of $[a, b]$, let $\Delta\alpha_j = \alpha(p_j) - \alpha(p_{j-1})$. Let f be a bounded function on $[a, b]$ and define *the upper Riemann sum* of f with respect to α and the *lower Riemann sum* of f with respect to α as follows:

$$\mathcal{U}(f, \mathcal{P}, \alpha) = \sum_{j=1}^{k} M_j \Delta\alpha_j$$

and

$$\mathcal{L}(f, \mathcal{P}, \alpha) = \sum_{j=1}^{k} m_j \Delta\alpha_j.$$

Here the notation M_j denotes the supremum of f on the interval $I_j = [p_{j-1}, p_j]$ and m_j denotes the infimum of f on I_j.

In the special case $\alpha(x) = x$ the Riemann sums discussed here have a form similar to the Riemann sums considered in the first two sections. Moreover,

$$\mathcal{L}(f, \mathcal{P}, \alpha) \le \mathcal{R}(f, \mathcal{P}) \le \mathcal{U}(f, \mathcal{P}, \alpha).$$

We define

$$I^*(f) = \inf \mathcal{U}(f, \mathcal{P}, \alpha)$$

and

$$I_*(f) = \sup \mathcal{L}(f, \mathcal{P}, \alpha).$$

Here the supremum and infimum are taken with respect to all partitions of the interval $[a, b]$. These are, respectively, the *upper* and *lower integrals* of f with respect to α on $[a, b]$.

By definition, it is always true that, for any partition $\mathcal{P}$,

$$\mathcal{L}(f, \mathcal{P}, \alpha) \le I_*(f) \le I^*(f) \le \mathcal{U}(f, \mathcal{P}, \alpha). \tag{$*$}$$

It is natural to declare the integral to exist when the upper and lower integrals agree:

DEFINITION 8.8 *Let α be a monotone increasing function on the interval $[a, b]$ and let f be a bounded function on $[a, b]$. We say that the **Riemann–Stieltjes integral of f with respect to** α exists if*

$$I^*(f) = I_*(f).$$

When the integral exists we denote it by

$$\int_a^b f \, d\alpha.$$

Notice that the definition of the Riemann–Stieltjes integral is different from the definition of Riemann integral that we used in the preceding sections. It turns out that when $\alpha(x) = x$ the two definitions are equivalent (this assertion is explored in the Exercises). In the present generality it is easier to deal with upper and lower integrals in order to determine the existence of integrals.

DEFINITION 8.9 *Let $\mathcal{P}$ and $\mathcal{Q}$ be partitions of the interval $[a, b]$. If each point of $\mathcal{P}$ is also an element of $\mathcal{Q}$ then we call $\mathcal{Q}$ a **refinement** of $\mathcal{P}$.*

Notice that the refinement $\mathcal{Q}$ is obtained by adding points to $\mathcal{P}$. The mesh of $\mathcal{Q}$ will be less than or equal to that of $\mathcal{P}$. The following lemma enables us to deal effectively with our new language.

LEMMA 8.13
Let $\mathcal{P}$ be a partition of the interval $[a, b]$ and f a function on $[a, b]$. Fix a monotone increasing function α on $[a, b]$. If $\mathcal{Q}$ is a refinement of $\mathcal{P}$ then

$$\mathcal{U}(f, \mathcal{Q}, \alpha) \le \mathcal{U}(f, \mathcal{P}, \alpha)$$

and

$$\mathcal{L}(f, \mathcal{Q}, \alpha) \geq \mathcal{L}(f, \mathcal{P}, \alpha).$$

PROOF Since $\mathcal{Q}$ is a refinement of $\mathcal{P}$ it holds that any interval I_ℓ arising from $\mathcal{Q}$ is contained in some interval $J_{j(\ell)}$ arising from $\mathcal{P}$. Let M_{I_ℓ} be the supremum of f on I_ℓ and $M_{J_{j(\ell)}}$ the supremum of f on the interval $J_{j(\ell)}$. Then $M_{I_\ell} \leq M_{J_{j(\ell)}}$. We conclude that

$$\mathcal{U}(f, \mathcal{Q}, \alpha) = \sum_\ell M_{I_\ell} \Delta\alpha_\ell \leq \sum_\ell M_{J_{j(\ell)}} \Delta\alpha_\ell.$$

We rewrite the right-hand side as

$$\sum_j M_{J_j} \left(\sum_{I_\ell \subseteq J_j} \Delta\alpha_\ell \right).$$

However, because α is monotone, the inner sum simply equals $\alpha(p_j) - \alpha(p_{j-1}) = \Delta\alpha_j$. Thus the last expression is equal to $\mathcal{U}(f, \mathcal{P}, \alpha)$, as desired.

A similar argument applies to the lower sums. ∎

Example 8.10

Let $[a, b] = [0, 10]$ and let $\alpha(x)$ be the *greatest integer function*. That is, $\alpha(x)$ is the greatest integer that does not exceed x. So, for example, $\alpha(0.5) = 0$, $\alpha(2) = 2$, and $\alpha(-3/2) = -2$. Certainly α is a monotone increasing function on $[0, 10]$. Let f be any continuous function on $[0, 10]$. We shall determine whether

$$\int_0^{10} f \, d\alpha$$

exists and, if it does, calculate its value.

Let $\mathcal{P}$ be a partition of $[0, 10]$. Assume for convenience that the points of the partition are distinct, and that none of these points (except the first and last) are integers. By the lemma, it is to our advantage to assume that the mesh of $\mathcal{P}$ is smaller than 1. Observe that $\Delta\alpha_j$ equals the number of integers that lie in the interval I_j — that is, either 0 or 1. Let $I_{j_0}, I_{j_2}, \ldots I_{j_{10}}$ be the intervals from the partition that do in fact contain integers (the first of these contains 0, the second contains 1, and so on up to 10). Then

$$\mathcal{U}(f, \mathcal{P}, \alpha) = \sum_{\ell=0}^{10} M_{j_\ell} \Delta\alpha_{j_\ell} = \sum_{\ell=1}^{10} M_{j_\ell}$$

and

$$\mathcal{L}(f, \mathcal{P}, \alpha) = \sum_{\ell=0}^{10} m_{j_\ell} \Delta\alpha_{j_\ell} = \sum_{\ell=1}^{10} m_{j_\ell}$$

because any term in these sums corresponding to an interval not containing an integer must have $\Delta\alpha_j = 0$. Notice that $\Delta\alpha_{j_0} = 0$ since $\alpha(0) = \alpha(p_1) = 0$.

Let $\epsilon > 0$. Since f is uniformly continuous on $[0, 10]$, we may choose a $\delta > 0$ such that $|s - t| < \delta$ implies that $|f(s) - f(t)| < \epsilon/20$. If $m(\mathcal{P}) < \delta$ then it follows that $|f(\ell) - M_{j_\ell}| < \epsilon/20$ and $|f(\ell) - m_{j_\ell}| < \epsilon/20$ for $\ell = 0, 1, \ldots, 10$. Therefore

$$\mathcal{U}(f, \mathcal{P}, \alpha) < \sum_{\ell=1}^{10} \left(f(\ell) + \frac{\epsilon}{20} \right)$$

and

$$\mathcal{L}(f, \mathcal{P}, \alpha) > \sum_{\ell=1}^{10} \left(f(\ell) - \frac{\epsilon}{20} \right).$$

Rearranging the first of these inequalities leads to

$$\mathcal{U}(f, \mathcal{P}, \alpha) < \left(\sum_{\ell=1}^{10} f(\ell) \right) + \frac{\epsilon}{2}$$

and, similarly,

$$\mathcal{L}(f, \mathcal{P}, \alpha) > \left(\sum_{\ell=1}^{10} f(\ell) \right) - \frac{\epsilon}{2}.$$

Thus, since I_* and I^* are trapped between $\mathcal{U}$ and $\mathcal{L}$, we conclude that

$$|I_*(f) - I^*(f)| < \epsilon.$$

We have seen that if the partition is fine enough then the upper and lower integrals of f with respect to α differ by at most ϵ. It follows that $\int_0^{10} f \, d\alpha$ exists. Moreover,

$$\left| I^*(f) - \sum_{\ell=1}^{10} f(\ell) \right| < \epsilon$$

and

$$\left| I_*(f) - \sum_{\ell=1}^{10} f(\ell) \right| < \epsilon.$$

We conclude that

$$\int_0^{10} f \, d\alpha = \sum_{\ell=1}^{10} f(\ell). \qquad \Box$$

The example demonstrates that the language of the Riemann–Stieltjes integral allows us to think of the integral as a generalization of the summation process. This is frequently useful, both philosophically and for practical reasons. The next result, sometimes called Riemann's Lemma, is crucial for proving the existence of Riemann–Stieltjes integrals.

PROPOSITION 8.14 RIEMANN'S LEMMA

Let α be a monotone increasing function on $[a, b]$ and f a bounded function on the interval. The Riemann–Stieltjes integral of f with respect to α exists if and only if for every $\epsilon > 0$ there is a partition $\mathcal{P}$ such that

$$|\mathcal{U}(f, \mathcal{P}, \alpha) - \mathcal{L}(f, \mathcal{P}, \alpha)| < \epsilon. \tag{$*$}$$

PROOF First assume that $(*)$ holds. Fix $\epsilon > 0$. Since $\mathcal{L} \leq I_* \leq I^* \leq \mathcal{U}$, inequality $(*)$ implies that

$$|I^*(f) - I_*(f)| < \epsilon.$$

But this means that $\int_a^b f \, d\alpha$ exists.

Conversely, assume that the integral exists. Fix $\epsilon > 0$. Choose a partition $\mathcal{Q}_1$ such that

$$|\mathcal{U}(f, \mathcal{Q}_1, \alpha) - I^*(f)| < \epsilon/2.$$

Likewise choose a partition $\mathcal{Q}_2$ such that

$$|\mathcal{L}(f, \mathcal{Q}_2, \alpha) - I_*(f)| < \epsilon/2.$$

Since $I_*(f) = I^*(f)$ it follows that

$$|\mathcal{U}(f, \mathcal{Q}_1, \alpha) - \mathcal{L}(f, \mathcal{Q}_2, \alpha)| < \epsilon. \tag{$**$}$$

Let $\mathcal{P}$ be the common refinement of $\mathcal{Q}_1$ and $\mathcal{Q}_2$. Then we have, again by Lemma 8.13, that

$$\mathcal{L}(f, \mathcal{Q}_2, \alpha) \leq \mathcal{L}(f, \mathcal{P}, \alpha) \leq \int_a^b f \, d\alpha \leq \mathcal{U}(f, \mathcal{P}, \alpha) \leq \mathcal{U}(f, \mathcal{Q}_1, \alpha).$$

But, by $(**)$, the expressions on the far left and the far right of these inequalities differ by less than ϵ. Thus $\mathcal{P}$ satisfies the condition $(*)$. ∎

We note in passing that the basic properties of the Riemann integral noted in Section 2 (Theorems 8.4 and 8.5) hold without change for the Riemann–Stieltjes integral. The proofs are left as exercises for you (use Riemann's Lemma!).

8.4   Advanced Results on Integration Theory

We now turn to establishing the existence of certain Riemann–Stieltjes integrals.

THEOREM 8.15
Let f be continuous on $[a, b]$ and assume that α is monotonically increasing. Then

$$\int_a^b f \, d\alpha$$

exists.

PROOF   We may assume that α is nonconstant; otherwise there is nothing to prove.

Pick $\epsilon > 0$. By the uniform continuity of f we may choose a $\delta > 0$ such that if $|s - t| < \delta$ then $|f(s) - f(t)| < \epsilon/(\alpha(b) - \alpha(a))$. Let $\mathcal{P}$ be any partition of $[a, b]$ that has mesh smaller than δ. Then

$$
\begin{aligned}
|\mathcal{U}(f, \mathcal{P}, \alpha) - \mathcal{L}(f, \mathcal{P}, \alpha)| &= \left| \sum_j M_j \Delta \alpha_j - \sum_j m_j \Delta \alpha_j \right| \\
&= \sum_j |M_j - m_j| \, \Delta \alpha_j \\
&\leq \sum_j \frac{\epsilon}{\alpha(b) - \alpha(a)} \Delta \alpha_j \\
&= \frac{\epsilon}{\alpha(b) - \alpha(a)} \cdot \sum_j \Delta \alpha_j \\
&= \epsilon.
\end{aligned}
$$

Here, of course, we have used the monotonicity of α to observe that the last sum collapses to $\alpha(b) - \alpha(a)$. By Riemann's Lemma, the proof is complete.
∎

Notice how simple Riemann's Lemma is to use. You may find it instructive to compare the proofs of this section with the rather difficult proofs in Section 2. What we are learning is that a good definition (and accompanying lemma(s)) can, in the end, make everything much simpler. Now we establish a companion result to the first one:

THEOREM 8.16
If α is a monotone increasing and continuous function on the interval $[a, b]$ and if f is monotonic on $[a, b]$ then $\int_a^b f \, d\alpha$ exists.

PROOF We may assume that $\alpha(b) > \alpha(a)$ and that f is monotone *increasing*. Let $L = \alpha(b) - \alpha(a)$ and $M = f(b) - f(a)$. Pick $\epsilon > 0$. Choose k a positive integer so that

$$\frac{L \cdot M}{k} < \epsilon.$$

Let $p_0 = a$ and choose p_1 to be the first point to the right of p_0 such that $\alpha(p_1) - \alpha(p_0) = L/k$ (this is possible, by the Intermediate Value Theorem, since α is continuous). Continuing, choose p_j to be the first point to the right of p_{j-1} such that $\alpha(p_j) - \alpha(p_{j-1}) = L/k$. This process will terminate after k steps and we will have $p_k = b$. Then $\mathcal{P} = \{p_0, p_1, \ldots, p_k\}$ is a partition of $[a, b]$.

Next observe that, for each j, the value M_j of sup f on I_j is $f(p_j)$ since f is monotone increasing. Similarly, the value m_j of inf f on I_j is $f(p_{j-1})$. We find therefore that

$$\mathcal{U}(f, \mathcal{P}, \alpha) - \mathcal{L}(f, \mathcal{P}, \alpha) = \sum_{j=1}^{k} M_j \Delta\alpha_j - \sum_{j=1}^{k} m_j \Delta\alpha_j$$

$$= \sum_{j=1}^{k} \left((M_j - m_j)\frac{L}{k} \right)$$

$$= \frac{L}{k} \sum_{j=1}^{k} \left(f(p_j) - f(p_{j-1}) \right)$$

$$= \frac{L \cdot M}{k}$$

$$< \epsilon.$$

Therefore inequality $(*)$ of Riemann's Lemma is satisfied and the integral exists. ∎

One of the useful features of Riemann–Stieltjes integration is that it puts integration by parts into a very natural setting. We begin with a lemma. In this lemma and the theorem following, we drop the standing hypothesis that our α be monotone increasing.

LEMMA 8.17
Let f be continuous on an interval $[a, b]$ and let g be continuous on that interval. If G is an antiderivative for g then $\int_a^b f \, dG$ exists and

$$\int_a^b f(x)g(x) \, dx = \int_a^b f \, dG.$$

PROOF Apply the Mean Value Theorem to the Riemann sums for the integral on the right. ∎

THEOREM 8.18   INTEGRATION BY PARTS
*Suppose that both f and g are continuous functions on the interval $[a, b]$. Let
F be an antiderivative for f on $[a, b]$ and G an antiderivative for g on $[a, b]$.
Then we have*

$$\int_a^b F\, dG = \left[F(b) \cdot G(b) - F(a) \cdot G(a) \right] - \int_a^b G\, dF.$$

PROOF   Notice that, by the preceding lemma, both integrals exist. Set $P(x) =
F(x) \cdot G(x)$. Then P has a continuous derivative on the interval $[a, b]$. Thus
the Fundamental Theorem applies and we may write

$$\int_a^b P'(x)\, dx = P(b) - P(a) = \left[F(b) \cdot G(b) - F(a) \cdot G(a) \right].$$

Now, writing out P' explicitly and using Leibnitz's Rule for the derivative of a
product, we obtain

$$\int_a^b F(x)g(x)\, dx = \left[F(b)G(b) - F(a)G(a) \right] - \int_a^b G(x)f(x)\, dx.$$

But the lemma allows us to rewrite this equation as

$$\int_a^b F\, dG = \left[F(b)G(b) - F(a)G(a) \right] - \int_a^b G(x)\, dF. \qquad\blacksquare$$

REMARK 8.19   The integration by parts formula can also be proved by applying
summation by parts to the Riemann sums for the integral

$$\int_a^b F\, dG.$$

This method is explored in the Exercises.   ∎

We have already observed that the Riemann–Stieltjes integral

$$\int_a^b f\, d\alpha$$

is linear in f; that is,

$$\int_a^b (f + g)\, d\alpha = \int_a^b f\, d\alpha + \int_a^b g\, d\alpha$$

and

$$\int_a^b c \cdot f\, d\alpha = c \cdot \int_a^b f\, d\alpha$$

when both f and g are Riemann–Stieltjes integrable with respect to α and for any constant c. We also would expect, from the very way that the integral is constructed, that it would be linear in the α entry. But we have not even defined the Riemann–Stieltjes integral for nonincreasing α. And what of a function α that is the difference of two monotone increasing functions? Such a function certainly need not be monotone. Is it possible to identify which functions α can be decomposed as sums or differences of monotonic functions? It turns out that there is a satisfactory answer to these questions, and we should like to discuss this matter briefly.

DEFINITION 8.11 *If α is a monotonically **decreasing** function on $[a, b]$ and f is a function on $[a, b]$ then we define*

$$\int_a^b f \, d\alpha = - \int_a^b f \, d(-\alpha)$$

when the right side exists.

The definition exploits the simple observation that if α is monotone decreasing then $-\alpha$ is monotone increasing; hence the preceding theory applies to the function $-\alpha$.

Next we have

DEFINITION 8.12 *Let α be a function on $[a, b]$ that can be expressed as*

$$\alpha(x) = \alpha_1(x) - \alpha_2(x),$$

where both α_1 and α_2 are monotone increasing. Then for any f on $[a, b]$ we define

$$\int_a^b f \, d\alpha = \int_a^b f \, d\alpha_1 - \int_a^b f \, d\alpha_2,$$

provided that both integrals on the right exist.

Now, by the very way that we have formulated our definitions, $\int_a^b f \, d\alpha$ is linear in both the f entry and the α entry. But the definitions are not satisfactory unless we can identify those α that can actually occur in the last definition. This leads us to a new class of functions.

DEFINITION 8.13 *Let f be a function on the interval $[a, b]$. For $x \in [a, b]$ we define*

$$Vf(x) = \sup \sum_{j=1}^k |f(p_j) - f(p_{j-1})|,$$

where the supremum is taken over all partitions $\mathcal{P}$ of the interval $[a, x]$.

*If $Vf \equiv Vf(b) < \infty$ then the function f is said to be of **bounded variation** on the interval $[a, b]$. In this circumstance the quantity $Vf(b)$ is called the **total variation** of f on $[a, b]$.*

A function of bounded variation has the property that its graph does not have unbounded total oscillation.

Example 8.14

Define $f(x) = \sin x$, with domain the interval $[0, 2\pi]$. Let us calculate Vf. Let $\mathcal{P}$ be a partition of $[0, 2\pi]$. Since adding points to the partition only makes the sum

$$\sum_{j=1}^{k} |f(p_j) - f(p_{j-1})|$$

larger (by the triangle inequality), we may as well suppose that $\mathcal{P} = \{p_0, p_1, p_2, \ldots, p_k\}$ contains the points $\pi/2, 3\pi/2$. Say that $p_{\ell_1} = \pi/2$ and $p_{\ell_2} = 3\pi/2$. Then

$$\sum_{j=1}^{k} |f(p_j) - f(p_{j-1})| = \sum_{j=1}^{\ell_1} |f(p_j) - f(p_{j-1})|$$
$$+ \sum_{j=\ell_1+1}^{\ell_2} |f(p_j) - f(p_{j-1})|$$
$$+ \sum_{j=\ell_2+1}^{k} |f(p_j) - f(p_{j-1})|.$$

However, f is monotone increasing on the interval $[0, \pi/2] = [0, p_{\ell_1}]$. Therefore the first sum is just

$$\sum_{j=1}^{\ell_1} f(p_j) - f(p_{j-1}) = f(p_{\ell_1}) - f(p_0) = f(\pi/2) - f(0) = 1.$$

Similarly, f is monotone on the intervals $[\pi/2, 3\pi/2] = [p_{\ell_1}, p_{\ell_2}]$ and $[3\pi/2, 2\pi] = [p_{\ell_2}, p_k]$. Thus the second and third sums equal $f(p_{\ell_1}) - f(p_{\ell_2}) = 2$ and $f(p_k) - f(p_{\ell_2}) = 1$ respectively. It follows that

$$Vf = Vf(2\pi) = 1 + 2 + 1 = 4.$$

Of course $Vf(x)$ for any $x \in [0, 2\pi]$ can be computed by similar means (see the Exercises). ⬜

In general, if f is a continuously differentiable function on an interval $[a, b]$ then

$$Vf(x) = \int_a^x |f'(t)| \, dt.$$

This assertion will be explored in the Exercises.

LEMMA 8.20
Let f be a function of bounded variation on the interval $[a, b]$. Then the function Vf is monotone increasing on $[a, b]$.

PROOF Let $s < t$ be elements of $[a, b]$. Let $\mathcal{P} = \{p_0, p_1, \ldots, p_k\}$ be a partition of $[a, s]$. Then $\tilde{\mathcal{P}} = \{p_0, p_1, \ldots, p_k, t\}$ is a partition of $[a, t]$ and

$$\sum_{j=1}^{k} |f(p_j) - f(p_{j-1})| \leq \sum_{j=1}^{k} |f(p_j) - f(p_{j-1})| + |f(t) - f(p_k)|$$

$$\leq Vf(t).$$

Taking the supremum on the left over all partitions $\mathcal{P}$ of $[a, s]$ yields that

$$Vf(s) \leq Vf(t). \qquad \blacksquare$$

LEMMA 8.21
Let f be a function of bounded variation on the interval $[a, b]$. Then the function $Vf - f$ is monotone increasing on the interval $[a, b]$.

PROOF Let $s < t$ be elements of $[a, b]$. Pick $\epsilon > 0$. By the definition of Vf we may choose a partition $\mathcal{P} = \{p_0, p_1, \ldots, p_k\}$ of the interval $[a, s]$ such that

$$Vf(s) - \epsilon < \sum_{j=1}^{k} |f(p_j) - f(p_{j-1})|. \qquad (*)$$

But then $\tilde{\mathcal{P}} = \{p_0, p_1, \ldots, p_k, t\}$ is a partition of $[a, t]$ and we have that

$$\sum_{j=1}^{k} |f(p_j) - f(p_{j-1})| + |f(t) - f(s)| \leq Vf(t).$$

Using $(*)$ we may conclude that

$$Vf(s) - \epsilon + f(t) - f(s) < \sum_{j=1}^{k} |f(p_j) - f(p_{j-1})| + |f(t) - f(s)| \leq Vf(t).$$

Therefore

$$Vf(s) - f(s) < Vf(t) - f(t) + \epsilon.$$

Since the inequality holds for every $\epsilon > 0$, we see that the function $Vf - f$ is monotone increasing. ∎

Now we may combine the last two lemmas to obtain our main result:

PROPOSITION 8.22

If a function f is of bounded variation on $[a, b]$, then f may be written as the difference of two monotone increasing functions. Conversely, the difference of two monotone increasing functions is a function of bounded variation.

PROOF   If f is of bounded variation write $f = Vf - (Vf - f) \equiv f_1 - f_2$. By the lemmas both f_1 and f_2 are monotone increasing.

For the converse, assume that $f = f_1 - f_2$ with f_1, f_2 monotone increasing. Then it is easy to see that

$$Vf(b) \leq |f_1(b) - f_1(a)| + |f_2(b) - f_2(a)| .$$

Thus f is of bounded variation. ∎

The main point of this discussion is the following theorem.

THEOREM 8.23

If f is a continuous function on $[a, b]$ and if α is of bounded variation on $[a, b]$ then the integral

$$\int_a^b f \, d\alpha$$

exists.

If g is of bounded variation on $[a, b]$ and if β is a continuous function of bounded variation on $[a, b]$ then the integral

$$\int_a^b g \, d\beta$$

exists.

PROOF   Write the function(s) of bounded variation as the difference of monotone increasing functions. Then apply Theorems 8.15 and 8.16. ∎

Exercises

8.1 If f is a Riemann integrable function on $[a, b]$ then show that f must be a bounded function.

8.2 Prove that if f is continuous on the interval $[a, b]$ except at finitely many points and is bounded then f is Riemann integrable on $[a, b]$.

8.3 Do Exercise 2 with the phrase "finitely many" replaced by "countably many."

8.4 Define the *Dirichlet function* to be

$$f(x) = \begin{cases} 1 & \text{if} \quad x \text{ is rational} \\ 0 & \text{if} \quad x \text{ is irrational.} \end{cases}$$

Prove that the Dirichlet function is not Riemann integrable on the interval $[0, 1]$.

8.5 Define

$$g(x) = \begin{cases} x \cdot \sin(1/x) & \text{if} \quad x \neq 0 \\ 0 & \text{if} \quad x = 0. \end{cases}$$

Is g Riemann integrable on the interval $[-1, 1]$?

8.6 Imitate the proof of the Fundamental Theorem of Calculus in Section 2 to show that if f is continuous on $[a, b]$ and if we define $F(x) = \int_a^x f(t)\, dt$ then $F'(a)$ exists and equals $f(a)$ in the sense that

$$\lim_{t \to a+} \frac{F(t) - F(a)}{t - a} = f(a).$$

Formulate and prove an analogous statement for the derivative of F at b.

8.7 Prove that if f is a continuously differentiable function on the interval $[a, b]$ then

$$Vf = \int_a^b |f'(x)|\, dx.$$

(Hint: You will prove two inequalities. For one, use the Fundamental Theorem. For the other, use the Mean Value Theorem.)

8.8 Provide the details of the assertion that if f is Riemann integrable on the interval $[a, b]$ then for any $\epsilon > 0$ there is a $\delta > 0$ such that if $\mathcal{P}$ is a partition of mesh less than δ then

$$\sum_j \left(\sup_{I_j} f - \inf_{I_j} f \right) \Delta_j < \epsilon.$$

(Hint: Follow the scheme presented before Definition 8.7. Given $\epsilon > 0$, choose δ as in the definition of the integral. Fix a partition $\mathcal{P}$ with mesh smaller than δ. Let $K + 1$ be the number of points in $\mathcal{P}$. Choose points $t_j \in I_j$ so that $|f(t_j) - \sup_{I_j} f| < \epsilon/(2(K + 1))$; also choose points $t'_j \in I_j$ so that $|f(t'_j) - \inf_{I_j} f| < \epsilon/(2(K + 1))$. By applying the definition of the integral to this choice of t_j and t'_j we find that

$$\sum_j \left(\sup_{I_j} f - \inf_{I_j} f \right) \Delta_j < 2\epsilon.$$

The result follows.)

8.9 Prove the converse of the statement in Exercise 8. (Hint: This is easier than Exercise 8, for any Riemann sum over a sufficiently fine partition $\mathcal{P}$ is trapped between the sum in which the infimum is always chosen and the sum in which the supremum is always chosen.)

8.10 Review the ideas in Exercises 8 and 9 as you verify that when $\alpha(x) = x$ then the Riemann–Stieltjes integral of a function f with respect to α on $[a, b]$ is just the same as the Riemann integral of f on $[a, b]$.

8.11 Let f be a bounded function on an unbounded interval of the form $[A, \infty)$. We say that f is integrable on $[A, \infty)$ if f is integrable on every compact subinterval of $[A, \infty)$ and

$$\lim_{B \to +\infty} \int_A^B f(x)\, dx$$

exists and is finite.

Assume that $f \geq 0$ is Riemann integrable on $[1, N]$ for every $N > 1$ and that f is monotone decreasing. Show that f is Riemann integrable on $[1, \infty)$ if and only if $\sum_{j=1}^{\infty} f(j)$ is finite.

Suppose that g is nonnegative and integrable on $[1, \infty)$. If $0 \leq |f(x)| \leq g(x)$ for $x \in [1, \infty)$ and f is integrable on compact subintervals of $[1, \infty)$ then prove that f is integrable on $[1, \infty)$.

8.12 Let f be a function on an interval of the form $(a, b]$ such that f is integrable on compact subintervals of $(a, b]$. If

$$\lim_{\epsilon \to 0^+} \int_{a+\epsilon}^b f(x)\, dx$$

exists and is finite then we say that f is integrable on $(a, b]$. Prove that in case we restrict attention to bounded f then in fact this definition gives rise to no new integrable functions. However there are unbounded functions that can now be integrated. Give an example.

Give an example of a function g that is integrable by the definition in the preceding paragraph but is such that $|g|$ is not integrable.

8.13 Prove that the integral

$$\int_0^\infty \frac{\sin x}{x}\, dx$$

exists by considering the limits of integrals $\int_\epsilon^N\, dx$ when $\epsilon \to 0^+$, $N \to \infty$. You may find it convenient to divide the integral.

8.14 State and prove the analogue of Theorem 8.4 for the Riemann–Stieltjes integral.

8.15 State and prove an analogue of Lemma 8.2 for the Riemann–Stieltjes integral.

8.16 Give an example to show that the composition of Riemann integrable functions need not be Riemann integrable.

8.17 Suppose that f is a continuous, nonnegative function on the interval $[0, 1]$. Let M be the supremum of f on the interval. Prove that

$$\lim_{n \to \infty} \left[\int_0^1 f(t)^n\, dt \right]^{1/n} = M.$$

8.18 Let f be a continuous function on the interval $[0, 1]$ that only takes nonnegative values there. Prove that

$$\left[\int_0^1 f(t)\, dt \right]^2 \leq \int_0^1 f(t)^2\, dt.$$

8.19 Let $f(x) = \sin x$ on the interval $[0, 2\pi]$. Calculate $Vf(x)$ for any $x \in [0, 2\pi]$.

8.20 Define $\alpha(x)$ by the condition that $\alpha(x) = -x + k$ when $k \leq x < k+1$. Calculate
$$\int_2^7 t^2 \, d\alpha(t).$$

8.21 Let $[x]$ be the greatest integer function as discussed in the text. Define the "fractional part" function by the formula $\alpha(x) = x - [x]$. Explain why this function has the name "fractional part." Calculate
$$\int_0^5 x \, d\alpha.$$

8.22 Give an example of a continuous function on the interval $[0,1]$ that is not of bounded variation.

8.23 To what extent is the following statement true? If f is Riemann integrable on $[a,b]$ then $1/f$ is Riemann integrable on $[a,b]$.

8.24 Explain how the summation by parts formula may be derived from the integration by parts formula proved in Section 4.

8.25 Explain how the integration by parts formula may be derived from the summation by parts process.

8.26 Let β be a monotone increasing function on the interval $[a,b]$. Set $m = \beta(a)$ and $M = \beta(b)$. For any number λ lying between m and M set $S_\lambda = \{x \in [a,b] : \beta(x) > \lambda\}$. Prove that S_λ must be an interval. Let $\ell(\lambda)$ be the length of S_λ. Then prove that
$$\int_a^b \beta(t)^p \, dt = -\int_m^M s^p \, d\ell(s)$$
$$= \int_0^M \ell(s) \cdot p \cdot s^{p-1} \, ds.$$

8.27 Give an example of a function f such that f^2 is Riemann integrable but f is not. What additional hypothesis on f would make the implication true?

8.28 Let f be a continuously differentiable function on the interval $[0, 2\pi]$. Further assume that $f(0) = f(2\pi)$ and $f'(0) = f'(2\pi)$. For $n \in \mathbb{N}$ define
$$\hat{f}(n) = \frac{1}{2\pi} \int_0^{2\pi} f(x) \sin nx \, dx.$$
Prove that
$$\sum_{n=1}^\infty |\hat{f}(n)|^2$$
converges. (Hint: Use integration by parts to obtain a favorable estimate on $|\hat{f}(n)|$.)

8.29 Prove that
$$\lim_{\eta \to 0^+} \int_\eta^{1/\eta} \frac{\cos(2r) - \cos r}{r} \, dr$$
exists.

8.30 If f is Riemann integrable on the interval $[a,b]$ and if $\mu : [\alpha, \beta] \to [a,b]$ is continuous then prove that $f \circ \mu$ is Riemann integrable on $[\alpha, \beta]$.

8.31 Prove that if $g(x)$ is continuous on $[a,b]$ and G is an antiderivative for g then there is a constant λ such that $G(x) + \lambda x$ is monotone increasing. Explain why this makes the sentence before Lemma 8.17 unnecessary.

9

Sequences and Series of Functions

9.1   Partial Sums and Pointwise Convergence

A *sequence of functions* is usually written

$$f_1(x), f_2(x), \ldots \quad \text{or} \quad \{f_j\}_{j=1}^{\infty} \, .$$

We will generally assume that the functions f_j all have the same domain S.

DEFINITION 9.1   *A sequence of functions* $\{f_j\}_{j=1}^{\infty}$ *with domain* $S \subseteq \mathbb{R}$ *is said to* **converge pointwise** *to a limit function* f *on* S *if for each* $x \in S$ *the sequence of numbers* $\{f_j(x)\}$ *converges to* $f(x)$.

Example 9.2
Define $f_j(x) = x^j$ with domain $S = \{x : 0 \leq x \leq 1\}$. If $0 \leq x < 1$ then $f_j(x) \to 0$. However $f_j(1) \to 1$. Therefore the sequence f_j converges to the function

$$f(x) = \begin{cases} 0 & \text{if} \quad 0 \leq x < 1 \\ 1 & \text{if} \quad x = 1. \end{cases} \qquad \square$$

Here are some of the basic questions that we must ask about a sequence of functions f_j that converges to a function f on a domain S:

1. If the functions f_j are continuous, then is f continuous? no
2. If the functions f_j are integrable on an interval I then is f integrable on I? If f is integrable on I, then does the sequence $\int_I f_j(x)\, dx$ converge to $\int_I f(x)\, dx$?
3. If the functions f_j are differentiable then is f differentiable? If f is differentiable then does the sequence f_j' converge to f'?

We see from Example 9.2 that the answer to the first question is "no": Each of the f_j is continuous but f certainly is not. It turns out that in order to obtain a favorable answer to our questions we must consider a stricter notion of convergence of functions. This motivates the next definition.

DEFINITION 9.3 *Let f_j be a sequence of functions on a domain S. We say that the functions f_j converge **uniformly** to f if, given $\epsilon > 0$, there is an $N > 0$ such that for any $j > N$ and any $x \in S$ it holds that $|f_j(x) - f(x)| < \epsilon$.*

Notice that the special feature of uniform convergence is that the rate at which $f_j(x)$ converges is independent of $x \in S$. In Example 9.2, $f_j(x)$ is converging very rapidly to zero for x near zero but arbitrarily slowly to zero for x near 1 (draw a sketch to help you understand this point). In the next example we shall prove this assertion rigorously:

Example 9.4

The sequence $f_j(x) = x^j$ does not converge uniformly to the limit function

$$f(x) = \begin{cases} 0 & \text{if} \quad 0 \le x < 1 \\ 1 & \text{if} \quad x = 1 \end{cases}$$

on the domain $S = [0, 1]$. In fact, it does not even do so on the smaller domain $(0, 1)$. To see this, notice that no matter how large j is, we have by the Mean Value Theorem that

$$f_j(1) - f_j(1 - 1/(2j)) = \frac{1}{2j} \cdot f_j'(\xi)$$

for some ξ between $1 - 1/(2j)$ and 1. But $f_j'(x) = j \cdot x^{j-1}$, hence $|f_j'(\xi)| < j$ and we conclude that

$$|f_j(1) - f_j(1 - 1/(2j))| < \frac{1}{2}$$

or

$$f_j(1 - 1/(2j)) > f_j(1) - \frac{1}{2} = \frac{1}{2}.$$

In conclusion, no matter how large j is, there will be a value of x (namely $x = 1 - 1/(2j)$) at which $f_j(x)$ is at least distance $1/2$ from the limit 0. We conclude that the convergence is not uniform. □

THEOREM 9.1

If f_j are continuous functions on a set S that converge uniformly on S to a function f then f is also continuous.

PROOF Let $\epsilon > 0$. Choose an integer N so large that if $j \geq N$ then $|f_j(x) - f(x)| < \epsilon/3$ for all $x \in S$. Fix $P \in S$. Choose $\delta > 0$ so small that if $|x - P| < \delta$ then $|f_N(x) - f_N(P)| < \epsilon/3$. For such x we have

$$|f(x) - f(P)| \leq |f(x) - f_N(x)| + |f_N(x) - f_N(P)| + |f_N(P) - f(P)|$$
$$< \frac{\epsilon}{3} + \frac{\epsilon}{3} + \frac{\epsilon}{3}$$

by the way that we chose N and δ. But the last line sums to ϵ, proving that f is continuous at P. Since $P \in S$ was chosen arbitrarily, we are done. ∎

Example 9.5

Define functions

$$f_j(x) = \begin{cases} 0 & \text{if} \quad x = 0 \\ j & \text{if} \quad 0 < x \leq 1/j \\ 0 & \text{if} \quad 1/j < x \leq 1. \end{cases}$$

Then $\lim_{j \to \infty} f_j(x) = 0$ for all x in the interval $I = [0, 1]$. However,

$$\int_0^1 f_j(x)\, dx = \int_0^{1/j} j\, dx = 1$$

for every j. Thus the f_j converge to the integrable limit function $f(x) \equiv 0$, but their integrals do not converge to the integral of f. □

Example 9.6

Let $q_1, q_2, \ldots$ be an enumeration of the rationals in the interval $I = [0, 1]$. Define functions

$$f_j(x) = \begin{cases} 1 & \text{if} \quad x \in \{q_1, q_2, \ldots, q_j\} \\ 0 & \text{if} \quad x \notin \{q_1, q_2, \ldots, q_j\}. \end{cases}$$

Then the functions f_j converge pointwise to the Dirichlet function f, which is equal to 1 on the rationals and 0 on the irrationals. Each of the functions f_j has integral 0 on I. But the function f is not integrable on I. □

The last two examples show that something more than pointwise convergence is needed in order for the integral to respect the limit process.

THEOREM 9.2

Let f_j be integrable functions on a bounded interval $[a, b]$ and suppose that the functions f_j converge uniformly to the limit function f. Then f is integrable on

$[a, b]$ *and*

$$\lim_{j \to \infty} \int_a^b f_j(x)\,dx = \int_a^b f(x)\,dx.$$

PROOF Assume that $a < b$. Pick $\epsilon > 0$. Choose N so large that if $j \geq N$ then $|f_j(x) - f(x)| < \epsilon/[2(b-a)]$ for all $x \in [a, b]$. Notice that

$$\left| \int_a^b f_j(x)\,dx - \int_a^b f_k(x)\,dx \right| \leq \int_a^b |f_j(x) - f_k(x)|\,dx. \qquad (*)$$

But if $j, k \geq N$ then $|f_j(x) - f_k(x)| \leq |f_j(x) - f(x)| + |f(x) - f_k(x)| < \epsilon/(b-a)$. Therefore line $(*)$ does not exceed

$$\int_a^b \frac{\epsilon}{b-a}\,dx = \epsilon.$$

Thus the numbers $\int_a^b f_j(x)\,dx$ form a Cauchy sequence. Let the limit of this sequence be called A. Notice that if we let $k \to \infty$ in the inequality

$$\left| \int_a^b f_j(x)\,dx - \int_a^b f_k(x)\,dx \right| \leq \epsilon$$

then we obtain

$$\left| \int_a^b f_j(x)\,dx - A \right| \leq \epsilon$$

for all $j \geq N$. This estimate will be used below.

By hypothesis there is a $\delta > 0$ such that if $\mathcal{P} = \{p_1, \ldots, p_k\}$ is a partition of $[a, b]$ with $m(\mathcal{P}) < \delta$ then

$$\left| \mathcal{R}(f_N, \mathcal{P}) - \int_a^b f_N(x)\,dx \right| < \epsilon.$$

But then for such a partition we have

$$|\mathcal{R}(f, \mathcal{P}) - A| \leq |\mathcal{R}(f, \mathcal{P}) - \mathcal{R}(f_N, \mathcal{P})| + \left| \mathcal{R}(f_N, \mathcal{P}) - \int_a^b f_N(x)\,dx \right|$$

$$+ \left| \int_a^b f_N(x)\,dx - A \right|.$$

We have already noted that, by the choice of N, the third term on the right is less than or equal to ϵ. The second term is smaller than ϵ by the way that we chose the partition $\mathcal{P}$. It remains to examine the first term. Now

$$|\mathcal{R}(f,\mathcal{P}) - \mathcal{R}(f_N,\mathcal{P})| = \left| \sum_{j=1}^{k} f(s_j)\Delta_j - \sum_{j=1}^{k} f_N(s_j)\Delta_j \right|$$

$$\leq \sum_{j=1}^{k} |f(s_j) - f_N(s_j)| \, \Delta_j$$

$$< \sum_{j=1}^{k} \frac{\epsilon}{2(b-a)} \Delta_j$$

$$= \frac{\epsilon}{2(b-a)} \sum_{j=1}^{k} \Delta_j$$

$$= \frac{\epsilon}{2}.$$

Therefore $|\mathcal{R}(f,\mathcal{P}) - A| < 3\epsilon$ when $m(\mathcal{P}) < \delta$. This shows that the function f is integrable on $[a,b]$ and has integral with value A. ∎

We have succeeded in answering questions (1) and (2), which were raised at the beginning of the section. In the next section we will answer question (3).

9.2 More on Uniform Convergence

In general, limits do not commute. Since the integral is defined with a limit, and since we saw in the last section that integrals do not always respect limits of functions, we know some concrete instances of noncommutation of limits. The fact that continuity is defined with a limit, and that the limit of continuous functions need not be continuous, gives even more examples of situations in which limits do not commute. Let us now turn to a situation in which limits *do* commute:

THEOREM 9.3

Fix a set S and a point $s \in S$. Assume that the functions f_j converge uniformly on the domain $S \setminus \{s\}$ to a limit function f. Suppose that each function $f_j(x)$ has a limit as $x \to s$. Then f itself has a limit as $x \to s$ and

$$\lim_{x \to s} f(x) = \lim_{j \to \infty} \lim_{x \to s} f_j(x).$$

Because of the way that f is defined, we may rewrite this conclusion as

$$\lim_{x \to s} \lim_{j \to \infty} f_j(x) = \lim_{j \to \infty} \lim_{x \to s} f_j(x).$$

In other words, the limits $\lim_{x \to s}$ and $\lim_{j \to \infty}$ commute.

PROOF   Let $\alpha_j = \lim_{x \to s} f_j(x)$. Let $\epsilon > 0$. There is a number $N > 0$ (independent of $x \in S \setminus \{s\}$) such that $j \geq N$ implies that $|f_j(x) - f(x)| < \epsilon/4$ for $x \in S \setminus \{s\}$. Fix $j, k \geq N$. Choose $\delta > 0$ such that $0 < |x - s| < \delta$ implies that $|f_j(x) - \alpha_j| < \epsilon/4$. Then for such an x we have

$$|\alpha_j - \alpha_k| \leq |\alpha_j - f_j(x)| + |f_j(x) - f(x)| + |f(x) - f_k(x)| + |f_k(x) - \alpha_k|.$$

The first and last expressions are less than $\epsilon/4$ by the choice of x. The middle two expressions are less than $\epsilon/4$ by the choice of N. We conclude that the sequence α_j is Cauchy. Let α be the limit of that sequence.

Letting $k \to \infty$ in the inequality

$$|\alpha_j - \alpha_k| < \epsilon$$

that we obtained above yields

$$|\alpha_j - \alpha| \leq \epsilon$$

for $j \geq N$. Now with δ as above and $0 < |x - s| < \delta$ we have

$$|f(x) - \alpha| \leq |f(x) - f_j(x)| + |f_j(x) - \alpha_j| + |\alpha_j - \alpha|.$$

By the choices we have made, the first term is less than $\epsilon/4$, the second is less than $\epsilon/4$, and the third is less than or equal to ϵ. Altogether, if $0 < |x - s| < \delta$ then $|f(x) - \alpha| < 2\epsilon$. This is the desired conclusion.   ∎

Parallel with our notion of Cauchy sequence of numbers, we have a concept of Cauchy sequence of functions in the uniform sense:

DEFINITION 9.7   *A sequence of functions f_j on a domain S is called **a uniformly Cauchy sequence** if for each $\epsilon > 0$ there is an $N > 0$ such that if $j, k > N$ then*

$$|f_j(x) - f_k(x)| < \epsilon \quad \forall x \in S.$$

PROPOSITION 9.4

A sequence of functions f_j is uniformly Cauchy on a domain S if and only if the sequence converges uniformly to a limit function f on the domain S.

PROOF   The proof is straightforward and is assigned as an exercise.   ∎

We will use the last two results in our study of the limits of differentiable functions. First we consider an example.

Example 9.8
Define the function

$$f_j(x) = \begin{cases} 0 & \text{if} \quad x \le 0 \\ jx^2 & \text{if} \quad 0 < x \le 1/(2j) \\ x - 1/(4j) & \text{if} \quad 1/(2j) < x < \infty. \end{cases}$$

We leave it as an exercise to check that the functions f_j converge uniformly on the entire real line to the function

$$f(x) = \begin{cases} 0 & \text{if} \quad x \le 0 \\ x & \text{if} \quad x > 0 \end{cases}$$

(draw a sketch to help you see this). Notice that each of the functions f_j is continuously differentiable on the entire real line, but f is not differentiable at 0.
￯

It turns out that we must strengthen our convergence hypotheses if we want the limit process to respect differentiation. The basic result is

THEOREM 9.5
*Suppose that a sequence f_j of differentiable functions on an open interval I converges pointwise to a limit function f. Suppose further that the sequence f_j' converges **uniformly** on I to a limit function g. Then the limit function f is differentiable on I and $f'(x) = g(x)$ for all $x \in I$.*

PROOF There is no loss of generality to assume that I is an interval of length 1. Let $\epsilon > 0$. The sequence $\{f_j'\}$ is uniformly Cauchy. Therefore we may choose N so large that $j, k > N$ implies that

$$\left| f_j'(x) - f_k'(x) \right| < \epsilon \quad \forall x \in I. \tag{$*$}$$

Fix a point $P \in I$. Define

$$\mu_j(x) = \frac{f_j(x) - f_j(P)}{x - P}$$

for $x \in I$, $x \ne P$. It is our intention to apply Theorem 9.3 to the functions μ_j.
    First notice that for each j we have

$$\lim_{x \to P} \mu_j(x) = f_j'(P).$$

Thus

$$\lim_{j \to \infty} \lim_{x \to P} \mu_j(x) = \lim_{j \to \infty} f_j'(P) = g(P).$$

That calculates the limits in one order. On the other hand,

$$\lim_{j \to \infty} \mu_j(x) = \frac{f(x) - f(P)}{x - P} \equiv \mu(x)$$

for $x \in I \setminus \{P\}$. If we can show that this convergence is uniform then Theorem 9.3 applies and we may conclude that

$$\lim_{x \to P} \mu(x) = \lim_{j \to \infty} \lim_{x \to P} \mu_j(x) = \lim_{j \to \infty} f_j'(P) = g(P).$$

But this just says that f is differentiable at P and the derivative equals g. That is the desired result.

To verify the uniform convergence of the μ_j, we apply the Mean Value Theorem to the function $f_j - f_k$. For $x \neq P$ we have

$$|\mu_j(x) - \mu_k(x)| = \frac{1}{|x - P|} \cdot |(f_j(x) - f_k(x)) - (f_j(P) - f_k(P))|$$

$$= \frac{1}{|x - P|} \cdot |x - P| \cdot |(f_j - f_k)'(\xi)|$$

$$= |(f_j - f_k)'(\xi)|$$

for some ξ between x and P. But line $(*)$ guarantees that the last line does not exceed ϵ. That shows that the μ_j converge uniformly and concludes the proof. ∎

REMARK 9.6 A little additional effort shows that we need only assume in the theorem that the functions f_j converge at a single point x_0 in the domain. One of the exercises asks you to prove this assertion.

Notice further that if we make the additional assumption that each of the functions f_j' is continuous then the proof of the theorem becomes much easier. For then

$$f_j(x) = f_j(x_0) + \int_{x_0}^{x} f_j'(t)\, dt$$

by the Fundamental Theorem of Calculus. The hypothesis that the f_j' converge uniformly then implies, by Theorem 9.2, that the integrals converge to

$$\int_{x_0}^{x} g(t)\, dt.$$

The hypothesis that the functions f_j converge at x_0 then allows us to conclude that the sequence $f_j(x)$ converges for every x to $f(x)$ and

$$f(x) = f(x_0) + \int_{x_0}^{x} g(t)\, dt.$$

The Fundamental Theorem of Calculus then yields that $f' = g$ as desired. ∎

9.3   Series of Functions

DEFINITION 9.9   *The formal expression*

$$\sum_{j=1}^{\infty} f_j(x),$$

*where the f_j are functions on a common domain S, is called a **series of functions**. For $N = 1, 2, 3, \ldots$, the expression*

$$S_N(x) = \sum_{j=1}^{N} f_j(x) = f_1(x) + f_2(x) + \ldots + f_N(x)$$

*is called the Nth **partial sum** for the series. In case*

$$\lim_{N \to \infty} S_N(x)$$

*exists and is finite we say that the series **converges** at x. Otherwise we say that the series **diverges** at x.*

Notice that the question of convergence of a series of functions, which should be thought of as an *addition process*, reduces to a question about the *sequence* of partial sums. Sometimes, as in the next example, it is convenient to begin the series at some index other than $j = 1$.

Example 9.10
Consider the series

$$\sum_{j=0}^{\infty} x^j.$$

This is the geometric series from Proposition 4.7. It converges absolutely for $|x| < 1$ and diverges otherwise.

By the formula for the partial sums of a geometric series,

$$S_N(x) = \frac{1 - x^{N+1}}{1 - x}.$$

For $|x| < 1$ we see that

$$S_N(x) \to \frac{1}{1 - x}.$$ □

DEFINITION 9.11 *Let*

$$\sum_{j=1}^{\infty} f_j(x)$$

*be a series of functions on a domain S. If the partial sums $S_N(x)$ converge uniformly on S to a limit function $g(x)$ then we say that the series **converges uniformly** on S.*

Of course, all of our results about uniform convergence of *sequences* of functions translate, via the sequence of partial sums of a series, to results about uniformly convergent series of functions. For example,

(a) If f_j are continuous functions on a domain S and if the series

$$\sum_{j=1}^{\infty} f_j(x)$$

converges uniformly on S to a limit function f, then f is also continuous on S.

(b) If f_j are integrable functions on $[a, b]$ and if

$$\sum_{j=1}^{\infty} f_j(x)$$

converges uniformly on $[a, b]$ to a limit function f, then f is also integrable on $[a, b]$ and

$$\int_a^b f(x)\, dx = \sum_{j=1}^{\infty} \int_a^b f_j(x)\, dx.$$

You will be asked to provide details of these assertions, as well as a statement and proof of a result about derivatives of series, in the Exercises. Meanwhile we turn to an elegant test for uniform convergence that is due to Weierstrass.

THEOREM 9.7   THE WEIERSTRASS M-TEST
Let $\{f_j\}_{j=1}^{\infty}$ be functions on a common domain S. Assume that each $|f_j|$ is bounded on S by a constant M_j and that

$$\sum_{j=1}^{\infty} M_j < \infty.$$

Then the series

$$\sum_{j=1}^{\infty} f_j \qquad\qquad (*)$$

converges uniformly on the set S.

PROOF   By hypothesis, the sequence T_N of partial sums of the series $\sum_{j=1}^{\infty} M_j$ is Cauchy. Given $\epsilon > 0$ there is therefore a number K so large that $q > p > K$ implies that

$$\sum_{j=p+1}^{q} M_j = |T_q - T_p| < \epsilon.$$

We may conclude that the partial sums S_N of the original series $\sum f_j$ satisfy, for $q > p > K$,

$$|S_q(x) - S_p(x)| = \left| \sum_{j=p+1}^{q} f_j(x) \right| \leq \sum_{j=p+1}^{q} |f_j(x)| \leq \sum_{j=p+1}^{q} M_j < \epsilon.$$

Thus the partial sums $S_N(x)$ of the series $(*)$ are uniformly Cauchy. The series $(*)$ therefore converges uniformly. ∎

Example 9.12
Let us consider the series

$$\sum_{j=1}^{\infty} 2^{-j} \sin\left(2^j x\right).$$

The sine terms oscillate so erratically that it would be difficult to calculate partial sums for this series. However, noting that the jth summand $f_j(x) = 2^{-j} \sin(2^j x)$ is dominated in absolute value by 2^{-j}, we see that the Weierstrass M-Test applies to this series. We conclude that the series converges uniformly on the entire real line.

By property (a) of uniformly convergent series of continuous functions, which was noted above, we may conclude that the function f defined by our series is continuous. It is also 2π-periodic: $f(x + 2\pi) = f(x)$ for every x since this assertion is true for each summand. Since the continuous function f restricted to the compact interval $[0, 2\pi]$ is uniformly continuous (Theorem 6.15), we may conclude that f is uniformly continuous on the entire real line.

However, it turns out that f is nowhere differentiable. The proof of this assertion follows lines similar to the treatment of nowhere-differentiable functions in Theorem 7.3. The details will be covered in an exercise.   ☐

9.4   The Weierstrass Approximation Theorem

The name Weierstrass has occurred frequently in this chapter. In fact, Karl Weierstrass (1815–1897) revolutionized analysis with his examples and theorems. This section is devoted to one of his most striking results. We introduce it with a motivating discussion.

It is natural to wonder whether the standard functions of calculus — $\sin x$, $\cos x$, and e^x, for instance — are actually polynomials of some very high degree. Since polynomials are so much easier to understand than these transcendental functions, an affirmative answer to this question would certainly simplify mathematics. Of course, a moment's thought shows that this wish is impossible: a polynomial of degree k has at most k real roots. Since sine and cosine have infinitely many real roots, they cannot be polynomials. A polynomial of degree k has the property that if it is differentiated enough times (namely $k + 1$ times) then the derivative is zero. Since this is not the case for e^x, we conclude that e^x cannot be a polynomial. The Exercises discuss other means for distinguishing the familiar transcendental functions of calculus from polynomial functions.

In calculus, however, we learned of a formal procedure, called Taylor series, for associating polynomials with a given function f. In some instances these polynomials form a sequence that converges back to the original function. This might cause us to speculate that any reasonable function can be approximated in some fashion by polynomials. In fact, the theorem of Weierstrass gives a spectacular affirmation of this speculation:

THEOREM 9.8   THE WEIERSTRASS APPROXIMATION THEOREM
*Let f be a continuous function on an interval $[a, b]$. Then there is a sequence of polynomials $p_j(x)$ with the property that the sequence p_j converges **uniformly** on $[a, b]$ to f.*

In a few moments we shall prove this theorem in detail. Let us first consider some of its consequences. A restatement of the theorem would be that, given a continuous function f on $[a, b]$ and an $\epsilon > 0$, there is a polynomial p such that

$$|f(x) - p(x)| < \epsilon$$

for every $x \in [a, b]$. If one were programming a computer to calculate values of a fairly wild function f, the theorem guarantees that, up to a given degree of accuracy, one could use a polynomial instead (which would in fact be much easier for the computer to handle). Advanced techniques can even tell what degree of polynomial is needed to achieve a given degree of accuracy. The proof that we shall present also suggests how this might be done.

Let f be the Weierstrass nowhere-differentiable function. The theorem guarantees that, on any compact interval, f is the uniform limit of polynomials. Thus even the uniform limit of infinitely differentiable functions need not be differentiable — even at one point. This explains why the hypotheses of Theorem 9.5 needed to be so stringent.

We shall break up the proof of the Weierstrass Approximation Theorem into a sequence of lemmas.

LEMMA 9.9

Let ψ_j be a sequence of continuous functions on the interval $[-1,1]$ with the following properties: (i) $\psi_j(x) \geq 0$ for all x; (ii) $\int_{-1}^{1} \psi_j(x)\,dx = 1$ for each j; (iii) for any $\delta > 0$ we have

$$\lim_{j \to \infty} \int_{\delta \leq |x| \leq 1} \psi_j(x)\,dx = 0.$$

If f is a continuous function on the real line that is identically zero off the interval $[0,1]$, then the functions $f_j(x) = \int_{-1}^{1} \psi_j(t)f(x-t)\,dt$ converge uniformly on the interval $[0,1]$ to $f(x)$.

PROOF   By multiplying f by a constant we may assume that $\sup|f| = 1$. Let $\epsilon > 0$. Since f is uniformly continuous on the interval $[0,1]$ we may choose a $\delta > 0$ such that if $|x - t| < \delta$ then $|f(x) - f(t)| < \epsilon/2$. By property (iii) above we may choose an N so large that $j > N$ implies that $|\int_{\delta \leq |t| \leq 1} \psi_j(t)\,dt| < \epsilon/4$. Then for any $x \in [0,1]$ we have

$$|f_j(x) - f(x)| = \left| \int_{-1}^{1} \psi_j(t)f(x-t)\,dt - f(x) \right|$$

$$= \left| \int_{-1}^{1} \psi_j(t)f(x-t)\,dt - \int_{-1}^{1} \psi_j(t)f(x)\,dt \right|.$$

Notice that in the last line we have used fact (ii) about the functions ψ_j to multiply the term $f(x)$ by 1 in a clever way. Now we may combine the two integrals to find that the last line

$$= \left| \int_{-1}^{1} (f(x-t) - f(x))\psi_j(t)\,dt \right|$$

$$\leq \int_{-\delta}^{\delta} |f(x-t) - f(x)|\psi_j(t)\,dt + \int_{\delta \leq |t| \leq 1} |f(x-t) - f(x)|\psi_j(t)\,dt$$

$$= A + B.$$

To estimate term A, we recall that for $|t| < \delta$ we have $|f(x - t) - f(x)| < \epsilon/2$; hence

$$A \leq \int_{-\delta}^{\delta} \frac{\epsilon}{2}\psi_j(t)\,dt \leq \frac{\epsilon}{2} \cdot \int_{-1}^{1} \psi_j(t)\,dt = \frac{\epsilon}{2}.$$

For B we take $j > N$ and write

$$B \leq \int_{\delta \leq |t| \leq 1} 2 \cdot \sup|f| \cdot \psi_j(t)\,dt \leq 2 \cdot \int_{\delta \leq |t| \leq 1} \psi_j(t)\,dt$$

$$< 2 \cdot \frac{\epsilon}{4} = \frac{\epsilon}{2},$$

where in the last inequality we have used the choice of j. Adding together our estimates for A and B, and noting that these estimates are independent of the choice of x, yields the result. $\blacksquare$

LEMMA 9.10

Define $\psi_j(t) = k_j \cdot (1 - t^2)^j$ on $[-1, 1]$, where the positive constants k_j are chosen so that $\int_{-1}^{1} \psi_j(t)\, dt = 1$. Then the functions ψ_j satisfy the properties (i)–(iii) of the last lemma.

PROOF Of course property (ii) is true by design. Property (i) is obvious. In order to verify property (iii), we need to estimate the size of k_j.

Notice that

$$\int_{-1}^{1} (1 - t^2)^j \, dt = 2 \cdot \int_{0}^{1} (1 - t^2)^j \, dt$$

$$\geq 2 \cdot \int_{0}^{1/\sqrt{j}} (1 - t^2)^j \, dt$$

$$\geq 2 \cdot \int_{0}^{1/\sqrt{j}} (1 - jt^2) \, dt,$$

where we have used the binomial theorem. But this last integral is easily evaluated and equals $2/(3\sqrt{j})$. We conclude that

$$\int_{-1}^{1} (1 - t^2)^j \, dt > \frac{1}{\sqrt{j}}.$$

As a result, $k_j < \sqrt{j}$.

Now, to verify property (iii) of the lemma, we notice that for $\delta > 0$ fixed and $\delta \leq |t| \leq 1$ it holds that

$$|\psi_j(t)| \leq k_j \cdot (1 - \delta^2)^j \leq \sqrt{j} \cdot (1 - \delta^2)^j$$

and this expression tends to 0 as $j \to \infty$. Thus $\psi_j \to 0$ uniformly on $\{t : \delta \leq |t| \leq 1\}$. It follows that the ψ_j satisfy property (iii) of the lemma. $\blacksquare$

PROOF OF THE WEIERSTRASS APPROXIMATION THEOREM We may assume without loss of generality (just by changing coordinates) that f is a continuous function on the interval $[0, 1]$. After adding a linear function (which is a polynomial) to f, we may assume that $f(0) = f(1) = 0$. Thus f may be continued to be identically zero off the interval $[0, 1]$.

Let ψ_j be as in Lemma 9.10 and form f_j as in Lemma 9.9. Then we know that f_j converge uniformly on $[0,1]$ to f. Finally,

$$f_j(x) = \int_{-1}^{1} \psi_j(t) f(x-t)\, dt$$

$$= \int_{0}^{1} \psi_j(x-t) f(t)\, dt$$

$$= k_j \int_{0}^{1} (1 + (x-t)^2)^j f(t)\, dt.$$

But multiplying out the expression $(1 + (x-t)^2)^j$ in the integrand then shows that f_j is a polynomial of degree at most $2j$ in x. Thus we have constructed a sequence of polynomials f_j that converges uniformly to f on the interval $[0,1]$.

∎

Exercises

9.1 Prove that if a series of continuous functions converges uniformly then the sum function is also continuous.

9.2 Prove that if a series $\sum_{j=1}^{\infty} f_j$ of integrable functions on an interval $[a,b]$ is uniformly convergent on $[a,b]$ then the sum function f is integrable and

$$\int_{a}^{b} f(x)dx = \sum_{j=1}^{\infty} \int_{a}^{b} f_j(x)\, dx.$$

9.3 Formulate and prove a result about the derivative of the sum of a convergent series of differentiable functions.

9.4 Let $0 < \alpha < 1$. Prove that the series

$$\sum_{j=1}^{\infty} 2^{-j\alpha} \sin\left(2^j x\right)$$

defines a function f that is nowhere differentiable. To achieve this end, follow the scheme that was used to prove Theorem 7.3: (1) Fix x. (2) For h small, choose M such that 2^{-M} is approximately equal to $|h|$. (3) Break the series up into the sum from 1 to $M-1$, the single summand $j = M$, and the sum from $j = M+1$ to ∞. The middle term has very large Newton quotient and the first and last terms are relatively small.

9.5 Prove Dini's Theorem: If f_j are continuous functions on a compact set K, $f_1(x) \leq f_2(x) \leq \ldots$ for all $x \in K$, and the f_j converge to a continuous function f on K, then in fact the f_j converge *uniformly* to f on K.

9.6 Prove Proposition 9.4. Refer to the parallel result in Chapter 3 for some hints.

9.7 Prove the assertion made in Remark 9.6 that Theorem 9.5 is still true if the functions f_j are assumed to converge at just one point (and also that the derivatives f_j' converge uniformly).

9.8 A function is called "piecewise linear" if it is (i) continuous and (ii) its graph consists of finitely many linear segments. Prove that a continuous function on an interval $[a, b]$ is the uniform limit of a sequence of piecewise linear functions.

9.9 If a sequence of functions f_j on a domain $S \subseteq \mathbb{R}$ has the property that $f_j \to f$ uniformly on S, then does it follow that $(f_j)^2 \to f^2$ uniformly on S? What simple additional hypothesis will make your answer affirmative?

9.10 If $f_j \to f$ uniformly on a domain S and if f_j, f never vanish on S then does it follow that the functions $1/f_j$ converge uniformly to $1/f$ on S?

9.11 Use the concept of boundedness of a function to show that the functions $\sin x$ and $\cos x$ cannot be polynomials.

9.12 Prove that if p is any polynomial then there is an N large enough that $e^x > |p(x)|$ for $x > N$. Conclude that the function e^x is not a polynomial.

9.13 Find a way to prove that $\tan x$ and $\ln x$ are not polynomials.

9.14 Let f_j be a uniformly convergent sequence of functions on a common domain S. What would be suitable conditions on a function ϕ to guarantee that $\phi \circ f_j$ converges uniformly on S?

9.15 Use the Weierstrass Approximation Theorem and Mathematical Induction to prove that if f is k times continuously differentiable on an interval $[a, b]$, then there is a sequence of polynomials p_j with the property that

$$p_j \to f$$

uniformly on $[a, b]$,

$$p'_j \to f'$$

uniformly on $[a, b]$,

$$\cdots$$

$$p_j^{(k)} \to f^{(k)}$$

uniformly on $[a, b]$.

9.16 Let $a < b$ be real numbers. Call a function of the form

$$f(x) = \begin{cases} 1 & \text{if} \quad a \leq x \leq b \\ 0 & \text{if} \quad x < a \text{ or } x > b \end{cases}$$

a *characteristic function* for the interval $[a, b]$. Then a function of the form

$$g(x) = \sum_{j=1}^{k} a_j \cdot f_j(x),$$

with the f_j characteristic functions of intervals $[a_j, b_j]$, is called *simple*. Prove that any continuous function on an interval $[c, d]$ is the uniform limit of a sequence of simple functions. (Hint: The proof of this assertion is conceptually simple; do *not* imitate the proof of the Weierstrass Approximation Theorem.)

9.17 Prove that the series

$$\sum_{j=1}^{\infty} \frac{\sin jx}{j}$$

converges uniformly on compact intervals that do not contain odd multiples of $\pi/2$. (Hint: Sum by parts and the result will follow.)

9.18 If f is a continuous function on the interval $[a, b]$ and if

$$\int_a^b f(x) p(x) \, dx = 0$$

for every polynomial p then prove that f must be the zero function. (Hint: Use Weierstrass's Approximation Theorem.)

9.19 Prove that the sequence of functions $f_j(x) = \sin(jx)$ has no subsequence that converges at every x.

9.20 Construct a sequence of continuous functions $f_j(x)$ that has the property that $f_j(q)$ increases monotonically to $+\infty$ for each rational q but such that, at each irrational x, $|f_j(x)| \le 1$ for infinitely many j.

9.21 Suppose that the sequence $f_j(x)$ on the interval $[0, 1]$ satisfies $|f_j(s) - f_j(t)| \le |s - t|$ for all $s, t \in [0, 1]$ (this is called a *Lipschitz condition*). Further assume that the f_j converge pointwise to a limit function f on the interval $[0, 1]$. Prove that the sequence converges uniformly.

9.22 Let $\{f_j\}$ be a sequence of continuous functions on the real line. Suppose that the f_j converge uniformly to a function f. Prove that

$$\lim_{j \to \infty} f_j(x + 1/j) = f(x)$$

uniformly on any bounded interval.

Can any of these hypotheses be weakened?

9.23 Prove a comparison test for uniform convergence of series: if $0 \le f_j \le g_j$ and the series $\sum g_j$ converges uniformly, then so also does the series $\sum f_j$.

9.24 Show by giving an example that the converse of the Weierstrass M-Test is false.

9.25 Define a *trigonometric polynomial* to be a function of the form

$$\sum_{j=1}^{k} a_j \cdot \cos jx + \sum_{j=1}^{\ell} b_j \cdot \sin jx.$$

Prove a version of the Weierstrass Approximation Theorem on the interval $[0, 2\pi]$ for 2π-periodic continuous functions and with the phrase "trigonometric polynomial" replacing "polynomial." (Hint: Prove that

$$\sum_{\ell=-j}^{j} \left(1 - \frac{|\ell|}{j+1} \right) (\cos \ell t) = \frac{1}{j+1} \left(\frac{\sin \frac{j+1}{2} t}{\sin \frac{1}{2} t} \right)^2 .$$

Use these functions as the $\psi_j's$ in the proof of Weierstrass's theorem.)

10

Special Functions

10.1 Power Series

A series of the form

$$\sum_{j=0}^{\infty} a_j (x - c)^j$$

is called a *power series* expanded about the point c. Our first task is to determine the nature of the set on which a power series converges.

PROPOSITION 10.1
Assume that the power series

$$\sum_{j=0}^{\infty} a_j (x - c)^j$$

converges at the value $x = d$. Let $r = |d - c|$. Then the series converges uniformly and absolutely on compact subsets of $\mathcal{I} = \{x : |x - c| < r\}$.

PROOF We may take the compact subset of $\mathcal{I}$ to be $K = [c - s, c + s]$ for some number $0 < s < r$. For $x \in K$ it then holds that

$$\sum_{j=0}^{\infty} \left| a_j (x - c)^j \right| = \sum_{j=0}^{\infty} \left| a_j (d - c)^j \right| \cdot \left| \frac{x - c}{d - c} \right|^j .$$

In the sum on the right, the first expression in absolute values is bounded by some constant C (by the convergence hypothesis). The quotient in absolute values is majorized by $L = s/r < 1$. The series on the right is thus dominated by

$$\sum_{j=0}^{\infty} C \cdot L^j .$$

This geometric series converges. By the Weierstrass M-Test, the original series converges absolutely and uniformly on K. ∎

An immediate consequence of the proposition is that the set on which the power series

$$\sum_{j=0}^{\infty} a_j(x-c)^j$$

converges is an interval centered about c. We call this set the *interval of convergence*. The series will converge absolutely and uniformly on compact subsets of the interval of convergence. The *radius* of the interval of convergence (called the *radius of convergence*) is defined to be half its length. Whether convergence holds at the endpoints of the interval will depend on the particular series. Let us use the notation $\mathcal{C}$ to denote the open interval of convergence.

It happens that if a power series converges at either of the endpoints of its interval of convergence, then the convergence is uniform up to that endpoint. This is a consequence of Abel's partial summation test; details will be explored in the Exercises.

On the interval of convergence $\mathcal{C}$, the power series defines a function f. Such a function is said to be **real analytic**. More precisely, we have

DEFINITION 10.1 *A function f, with domain an open set $U \subseteq \mathbb{R}$ and range either the real or the complex numbers, is called **real analytic** if for each $c \in U$ the function f may be represented by a convergent power series on an interval of positive radius centered at c:*

$$f(x) = \sum_{j=0}^{\infty} a_j(x-c)^j.$$

We need to know both the algebraic and the calculus properties of a real analytic function: Is it continuous? differentiable? How does one add/subtract/ multiply/divide two such functions?

PROPOSITION 10.2

Let

$$\sum_{j=0}^{\infty} a_j(x-c)^j \quad and \quad \sum_{j=0}^{\infty} b_j(x-c)^j$$

be two power series with intervals of convergence $\mathcal{C}_1$ and $\mathcal{C}_2$ centered at c. Let $f_1(x)$ be the function defined by the first series on $\mathcal{C}_1$ and $f_2(x)$ the function

defined by the second series on $\mathcal{C}_2$. *Then on their common domain* $\mathcal{C} = \mathcal{C}_1 \cap \mathcal{C}_1$
it holds that

1. $f(x) \pm g(x) = \sum_{j=0}^{\infty} (a_j \pm b_j)(x - c)^j$;
2. $f(x) \cdot g(x) = \sum_{m=0}^{\infty} \sum_{j+k=m} (a_j \cdot b_k)(x - c)^m$.

PROOF Let

$$A_N = \sum_{j=0}^{N} a_j (x - c)^j \quad \text{and} \quad B_N = \sum_{j=0}^{N} b_j (x - c)^j$$

be, respectively, the Nth partial sums of the power series that define f and g.
If C_N is the Nth partial sum of the series

$$\sum_{j=0}^{\infty} (a_j \pm b_j)(x - c)^j$$

then

$$f(x) \pm g(x) = \lim_{N \to \infty} A_N \pm \lim_{N \to \infty} B_N = \lim_{N \to \infty} [A_N \pm B_N]$$

$$= \lim_{N \to \infty} C_N = \sum_{j=0}^{\infty} (a_j \pm b_j)(x - c)^j.$$

This proves (1).

For (2), we could just apply the theory of the Cauchy product. However, for
completeness, we repeat the key ideas.

$$D_N = \sum_{m=0}^{N} \sum_{j+k=m} (a_j \cdot b_k)(x - c)^m \quad \text{and} \quad R_N = \sum_{j=N+1}^{\infty} b_j (x - c)^j.$$

We have

$$D_N = a_0 B_N + a_1 (x - c) B_{N-1} + \ldots + a_N (x - c)^N B_0$$

$$= a_0 (g(x) - R_N) + a_1 (x - c)(g(x) - R_{N-1})$$

$$+ \ldots + a_N (x - c)^N (g(x) - R_0)$$

$$= g(x) \sum_{j=0}^{N} a_j (x - c)^j$$

$$- [a_0 R_N + a_1 (x - c) R_{N-1} + \ldots + a_N (x - c)^N R_0].$$

Clearly,

$$g(x) \sum_{j=0}^{N} a_j (x - c)^j$$

converges to $g(x)f(x)$ as N approaches ∞. In order to show that $D_N \to g \cdot f$, it will thus suffice to show that

$$\left| a_0 R_N + a_1(x-c)R_{N-1} + \ldots + a_N(x-c)^N R_0 \right|$$

converges to 0 as N approaches ∞. Fix x. Now we know that

$$\sum_{j=0}^{\infty} a_j (x-c)^j$$

is absolutely convergent so we may set

$$A = \sum_{j=0}^{\infty} |a_j||x-c|^j.$$

Also $\sum_{j=0}^{\infty} b_j(x-c)^j$ is convergent. Therefore, given $\epsilon > 0$, we can find N_0 so that $N \geq N_0$ implies $|R_N| \leq \epsilon$. Thus we have

$$\left| a_0 R_N + a_1(x-c)R_{N-1} + \ldots + a_N(x-c)^N R_0 \right|$$

$$\leq \left| a_0 R_N + \ldots + a_{N-N_0}(x-c)^{N-N_0} R_{N_0} \right|$$

$$+ \left| a_{N-N_0+1}(x-c)^{N-N_0+1} R_{N_0-1} \ldots + a_N(x-c)^N R_0 \right|$$

$$\leq \sup_{M \geq N_0} R_M \cdot \left(\sum_{j=0}^{\infty} |a_j||x-c|^j \right)$$

$$+ \left| a_{N-N_0+1}(x-c)^{N-N_0+1} R_{N_0-1} \ldots + a_N(x-c)^N R_0 \right|$$

$$\leq \epsilon A + \left| a_{N-N_0+1}(x-c)^{N-N_0+1} R_{N_0-1} \ldots + a_N(x-c)^N R_0 \right|.$$

Thus

$$\left| a_0 R_N + a_1(x-c)R_{N-1} + \cdots + a_N(x-c)^N R_0 \right|$$

$$\leq \epsilon \cdot A + M \cdot \sum_{j=N-N_0+1}^{N} |a_j||x-c|^j,$$

where M is an upper bound for $|R_j(x)|$. Since the series defining A converges, we find on letting $N \to \infty$ that

$$\limsup_{N \to \infty} \left| a_0 R_N + a_1(x-c)R_{N-1} + \cdots + a_N(x-c)^N R_0 \right| \leq \epsilon \cdot A.$$

Since $\epsilon > 0$ was arbitrary, we may conclude that

$$\lim_{N \to \infty} \left| a_0 R_N + a_1(x-c)R_{N-1} + \cdots + a_N(x-c)^N R_0 \right| = 0. \qquad \blacksquare$$

Next we turn to division of real analytic functions. If f and g are real analytic functions both defined on an open interval I, and if g does not vanish on I, then we would like f/g to be a well-defined real analytic function (it certainly is a well-defined *function*) and we would like to be able to calculate its power series expansion by formal long division. This is what the next result tells us:

PROPOSITION 10.3
Let f and g be real analytic functions, both of which are defined on an open interval I. Assume that g does not vanish on I. Then the function

$$h(x) = \frac{f(x)}{g(x)}$$

is real analytic on I. Moreover, if I is centered at the point c and if

$$f(x) = \sum_{j=0}^{\infty} a_j(x - c)^j \quad and \quad g(x) = \sum_{j=0}^{\infty} b_j(x - c)^j$$

then the power series expansion of h about c may be obtained by formal long division of the latter series into the former. That is, the zeroeth coefficient c_0 of h is

$$c_0 = a_0/b_0,$$

the order one coefficient c_1 is

$$c_1 = \frac{1}{b_0}\left(a_1 - \frac{a_0 b_1}{b_0}\right),$$

etc.

PROOF   If we can show that the power series

$$\sum_{j=0}^{\infty} c_j(x - c)^j$$

just indicated converges on I then the result on multiplication of series in Proposition 10.2 yields the result. There is no loss of generality in assuming that $c = 0$. Assume also that $b_1 \neq 0$.

Notice that one may check inductively that, for $j \geq 1$,

$$c_j = \frac{1}{b_0}\left(a_j - b_1 \cdot c_{j-1}\right). \tag{$*$}$$

Without loss of generality, we may scale the a_j's and the b_j's and assume that the radius of I is $1 + \epsilon$, some $\epsilon > 0$. Then we see from the last displayed formula that

$$|c_j| \leq C \cdot \left(|a_j| + |c_{j-1}|\right),$$

where $C = \max\{|1/b_0|, |b_1/b_0|\}$. It follows that

$$|c_j| \leq C' \cdot \left(1 + |a_j| + |a_{j-1}| + \cdots + |a_0|\right).$$

Since the radius of I exceeds 1, $\sum |a_j| < \infty$ and we see that the $|c_j|$ are bounded. Hence the power series with coefficients c_j has radius of convergence 1.

In case $b_1 = 0$ then the role of b_1 is played by the first nonvanishing b_m, $m > 1$. Then a new version of formula $(*)$ is obtained and the argument proceeds as before. ∎

In practice it is often useful to calculate f/g by expanding g in a "geometric series." To illustrate this idea, we assume for simplicity that f and g are real analytic in a neighborhood of 0. Then

$$\frac{f(x)}{g(x)} = f(x) \cdot \frac{1}{g(x)}$$

$$= f(x) \cdot \frac{1}{b_0 + b_1 x + \cdots}$$

$$= f(x) \cdot \frac{1}{b_0} \cdot \frac{1}{1 + (b_1/b_0)x + \cdots}.$$

Now we use the fact that, for β small,

$$\frac{1}{1 - \beta} = 1 + \beta + \beta^2 + \cdots.$$

Setting $\beta = -(b_1/b_0)x - \ldots$ and substituting the resulting expansion into our expression for $f(x)/g(x)$ then yields a formula that can be multiplied out to give a power series expansion for $f(x)/g(x)$.

10.2   More on Power Series: Convergence Issues

We now introduce the *Hadamard formula* for the radius of convergence of a power series.

LEMMA 10.4

For the power series

$$\sum_{j=0}^{\infty} a_j (x - c)^j$$

define A and ρ by

$$A = \limsup_{n \to \infty} |a_n|^{1/n},$$

$$\rho = \begin{cases} 0 & \text{if } A = \infty \\ 1/A & \text{if } 0 < A < \infty \\ \infty & \text{if } A = 0, \end{cases}$$

then ρ is the radius of convergence of the power series about c.

PROOF   Observing that

$$\limsup_{n \to \infty} |a_n(x-c)^n|^{1/n} = A|x-c|,$$

we see the lemma is an immediate consequence of the Root Test.   ∎

COROLLARY 10.5
The power series

$$\sum_{j=0}^{\infty} a_j(x-c)^j$$

has radius of convergence ρ if and only if, when $0 < R < \rho$, there exists a constant $0 < C = C_R$ such that

$$|a_n| \le \frac{C}{R^n}.$$

From the power series

$$\sum_{j=0}^{\infty} a_j(x-c)^j$$

it is natural to create the *derived series*

$$\sum_{j=1}^{\infty} j a_j(x-c)^j$$

using term-by-term differentiation.

PROPOSITION 10.6
The radius of convergence of the derived series is the same as the radius of convergence of the original power series.

PROOF   We observe that

$$\limsup_{n \to \infty} |na_n|^{1/n} = \lim_{n \to \infty} n^{-1/n} \limsup_{n \to \infty} |na_n|^{1/n}$$

$$= \limsup_{n \to \infty} |a_n|^{1/n}.$$

So the result follows from the Hadamard formula.   ∎

PROPOSITION 10.7
Let f be a real analytic function defined on an open interval I. Then f is continuous and has continuous, real analytic derivatives of all orders. In fact, the derivatives of f are obtained by differentiating its series representation term by term.

PROOF   Since for each $c \in I$ the function f may be represented by a convergent power series with positive radius of convergence, we see that in a sufficiently small open interval about each $c \in I$, f is the uniform limit of a sequence of continuous functions: the partial sums of the power series representing f. It follows that f is continuous at c. Since the radius of convergence of the derived series is the same as that of the original series, it also follows that the derivatives of the partial sums converge uniformly on the same open interval about c to a continuous function. It then follows from Theorem 9.5 that f is differentiable and its derivative is the function defined by the derived series. By induction, f has continuous derivatives of all orders at c.   ∎

We can now show that a real analytic function has a unique power series representation at any point.

COROLLARY 10.8
If the function f is represented by a convergent power series on an interval of positive radius centered at c,

$$f(x) = \sum_{j=0}^{\infty} a_j (x - c)^j,$$

then the coefficients of the power series are related to the derivatives of the function by

$$a_j = \frac{f^{(j)}(c)}{j!}.$$

PROOF   This follows readily by differentiating both sides of the above equation j times, as we may by the proposition, and evaluating at $x = c$.   ∎

Finally, we note that integration of power series is as well behaved as differentiation.

PROPOSITION 10.9
The power series

$$\sum_{j=0}^{\infty} a_j (x - c)^j$$

and the series

$$\sum_{j=0}^{\infty} \frac{a_j}{j+1} (x - c)^{j+1}$$

obtained from term-by-term integration have the same radius of convergence. The function F defined by

$$F(x) = \sum_{j=0}^{\infty} \frac{a_j}{j+1} (x - c)^{j+1}$$

on the common interval of convergence satisfies

$$F'(x) = \sum_{j=0}^{\infty} a_j (x - c)^j.$$

PROOF   The proof is left to the Exercises.   ∎

It is sometimes convenient to allow the variable in a power series to be a complex number. In this case we write

$$\sum_{j=0}^{\infty} a_j (z - c)^j,$$

where z is the complex argument. We now allow c and the a_j's to be complex numbers as well. Noting that the elementary facts about series hold for complex series as well as real series (you should check this for yourself), we see that the arguments of this section show that the domain of convergence of a complex power series is a *disc* in the complex plane with radius ρ given as follows:

$$A = \limsup_{n \to \infty} |a_n|^{1/n}$$

$$\rho = \begin{cases} 0 & \text{if } A = \infty \\ 1/A & \text{if } 0 < A < \infty \\ \infty & \text{if } A = 0. \end{cases}$$

The proofs in this section apply to show that convergent complex power series may be added, subtracted, multiplied, and divided (provided that we do not divide by zero) on their common domains of convergence. They may also be differentiated and integrated term by term.

These observations about complex power series will be useful in the next section.

We conclude this section with a consideration of Taylor series:

THEOREM 10.10   *TAYLOR'S EXPANSION*
For k a nonnegative integer let f be a $k + 1$ times continuously differentiable function on an open interval $I = (a - \epsilon, a + \epsilon)$. Then, for $x \in I$,

$$f(x) = \sum_{j=0}^{k} f^{(j)}(a) \frac{(x - a)^j}{j!} + R_{k,a}(x),$$

where

$$R_{k,a}(x) = \int_{a}^{x} f^{(k+1)}(t) \frac{(x - t)^k}{k!} \, dt.$$

PROOF   We apply integration by parts to the Fundamental Theorem of Calculus to obtain

$$f(x) = f(a) + \int_{a}^{x} f'(t) \, dt$$

$$= f(a) + \left(f'(t) \frac{(t - x)}{1!} \right) \Big|_{a}^{x} - \int_{a}^{x} f''(t) \frac{(t - x)}{1!} \, dt$$

$$= f(a) + f'(a) \frac{(x - a)}{1!} + \int_{a}^{x} f''(t) \frac{x - t}{1!} \, dt.$$

Notice that when we performed the integration by parts we used $t - x$ as an antiderivative for dt. This is of course legitimate, as a glance at the integration by parts theorem reveals. We have proved the theorem for the case $k = 1$. The result for higher k is obtained inductively by repeated integrations by parts.   ∎

Taylor's theorem allows us to associate with any infinitely differentiable function a formal expansion of the form

$$\sum_{j=0}^{\infty} a_j (x - a)^j.$$

However, there is no guarantee that this series will converge; even if it does converge, it may not converge back to $f(x)$. An important example to keep in mind is the function

$$h(x) = \begin{cases} 0 & \text{if} \quad x = 0 \\ e^{-1/x^2} & \text{if} \quad x \neq 0. \end{cases}$$

This function is infinitely differentiable at every point of the real line (including 0). However, all of its derivatives at $x = 0$ are equal to zero (this matter will be treated in the Exercises). Therefore the formal Taylor series expansion of h about $a = 0$ is

$$\sum_{j=0}^{\infty} 0 \cdot (x - 0)^j = 0.$$

We see that the formal Taylor series expansion for h converges to the zero function at every x, but not to the original function h itself.

In fact, the theorem tells us that the Taylor expansion of a function f converges to f at a point x if and only if $R_{k,a}(x) \to 0$. In the Exercises we shall explore the following more quantitative assertion:

> An infinitely differentiable function f on an interval I has Taylor series expansion about $a \in I$ that converges back to f on a neighborhood J of a if and only if there are positive constants C, R such that for every $x \in J$ and every k it holds that
>
> $$\left| f^{(k)}(x) \right| \leq C \cdot \frac{k!}{R^k}.$$

The function h considered above should not be thought of as an isolated exception. For instance, we know from calculus that the function $f(x) = \sin x$ has Taylor expansion that converges to f at every x. But then for ϵ small the function $g_\epsilon(x) = f(x) + \epsilon \cdot h(x)$ has Taylor series that does *not* converge back to $g_\epsilon(x)$ for $x \neq 0$. Similar examples may be generated by using other real analytic functions in place of sine.

10.3 The Exponential and Trigonometric Functions

We begin by defining the exponential function:

DEFINITION 10.2 *The power series*

$$\sum_{j=0}^{\infty} \frac{z^j}{j!}$$

*converges, by the Ratio Test, for every complex value of z. The function defined thereby is called the **exponential function** and is written $\exp(z)$.*

PROPOSITION 10.11
The function $\exp(z)$ *satisfies*

$$\exp(a + b) = \exp(a) \cdot \exp(b)$$

for any complex numbers a and b.

PROOF   We write the right-hand side as

$$\sum_{j=0}^{\infty} \frac{a^j}{j!} \sum_{j=0}^{\infty} \frac{b^j}{j!}.$$

Now convergent power series may be multiplied term by term. We find that the last line equals

$$\sum_{j=0}^{\infty} \left(\sum_{\ell=0}^{j} \frac{a^{(j-\ell)}}{(j-\ell)!} \cdot \frac{b^\ell}{\ell!} \right). \tag{$*$}$$

However, the inner sum on the right side of this equation may be written as

$$\frac{1}{j!} \sum_{\ell=0}^{j} \frac{j!}{\ell!(j-\ell)!} a^{j-\ell} b^\ell = \frac{1}{j!} (a+b)^j.$$

It follows that line $(*)$ equals $\exp(a + b)$.   ∎

We set $e = \exp(1)$. This is consistent with our earlier treatment of the number e in Section 4.4. The proposition tells us that for any positive integer k we have

$$e^k = e \cdot e \cdots e = \exp(1) \cdot \exp(1) \cdots \exp(1) = \exp(k).$$

If m is another positive integer then

$$\big(\exp(k/m)\big)^m = \exp(k) = e^k$$

whence

$$\exp(k/m) = e^{k/m}.$$

We may extend this formula to *negative* rational exponents by using the fact that $\exp(a) \cdot \exp(-a) = 1$. Thus, for any rational number q,

$$\exp(q) = e^q.$$

Now note that the function exp is monotone increasing and continuous. It follows (this fact is treated in the Exercises) that if we set, for any $r \in \mathbb{R}$,

$$e^r = \sup\{e^q : q \in \mathbb{Q}, q < r\}$$

(this is a *definition* of the expression e^r) then $e^x = \exp(x)$ for every real x. [You may find it useful to review the discussion of exponentiation in Section 3.4; the presentation here parallels that one.] We will adhere to custom and write e^x instead of $\exp(x)$ when the argument of the function is real.

PROPOSITION 10.12
The exponential function e^x satisfies

1. $e^x > 0$ *for all x.*
2. $e^0 = 1$.
3. $(e^x)' = e^x$.
4. e^x *is strictly increasing.*
5. *The graph of e^x is asymptotic to the negative x-axis.*
6. *For each integer $N > 0$ there is a number $c_N > 0$ such that $e^x > c_N \cdot x^N$ when $x > 0$.*

PROOF The first three statements are obvious from the power series expansion for the exponential function.

If $s < t$ then the Mean Value Theorem tells us that there is a number ξ between s and t such that

$$e^t - e^s = (t - s) \cdot e^\xi > 0;$$

hence the exponential function is strictly increasing.

By inspecting the power series we see that $e^x > 1 + x$, hence e^x increases to $+\infty$. Since $e^x \cdot e^{-x} = 1$ we conclude that e^{-x} tends to 0 as $x \to +\infty$. Thus the graph of the exponential function is asymptotic to the negative x-axis.

Finally, by inspecting the power series for e^x we see that assertion 6 is true with $c_N = 1/N!$. ∎

Now we turn to the trigonometric functions. The definition of the trigonometric functions that is found in calculus texts is unsatisfactory because it relies too heavily on a picture and because the continual need to subtract off superfluous multiples of 2π is clumsy. We have nevertheless used the trigonometric functions in earlier chapters to illustrate various concepts. It is time now to give a rigorous definition of the trigonometric functions that is independent of these earlier considerations.

DEFINITION 10.3 *The power series*

$$\sum_{j=0}^{\infty} (-1)^j \frac{x^{2j+1}}{(2j+1)!}$$

converges at every point of the real line (by the Ratio Test). The function that it defines is called the sine *function and is usually written* sin x.

The power series

$$\sum_{j=0}^{\infty}(-1)^j\,\frac{x^{2j}}{(2j)!}$$

converges at every point of the real line (by the Ratio Test). The function that it defines is called the cosine *function and is usually written* cos x.

You may recall that the power series that we use to define the sine and cosine functions are precisely the Taylor series expansions for the functions sine and cosine that were derived in your calculus text. But now we *begin* with the power series and must derive the properties of sine and cosine that we need *from these series*.

In fact, the most convenient way to achieve this goal is to proceed by way of the exponential function. (The point here is mainly one of convenience. It can be verified by direct manipulation of the power series that $\sin^2 x + \cos^2 x = 1$ and so forth, but the algebra is extremely unpleasant.) The formula in the next proposition is usually credited to Euler.

PROPOSITION 10.13
The exponential function and the functions sine *and* cosine *are related by the formula (for x and y real and $i^2 = -1$)*

$$\exp(x + iy) = e^x \cdot (\cos y + i \sin y).$$

PROOF   One simply writes out the series on the right, multiplies them together term by term, and collects terms. The result follows.   ∎

Because of this formula, we know that

$$\exp(iy) = \cos y + i \sin y. \qquad\qquad (**)$$

We will usually write this as $e^{iy} = \cos y + i\sin y$, where this expression *defines* what we mean by e^{iy}. As a result,

$$e^{x+iy} = e^x \cdot e^{iy} = e^x \cdot (\cos y + i \sin y).$$

Notice that $e^{-iy} = \cos(-y) + i\sin(-y) = \cos y - i\sin y$ (we know from their power series expansions that the sine function is odd and the cosine function even). Then formula $(**)$ tells us that

$$\cos y = \frac{e^{iy} + e^{-iy}}{2}$$

and

$$\sin y = \frac{e^{iy} - e^{-iy}}{2i}.$$

Now we may prove:

PROPOSITION 10.14
For every real x it holds that

$$\sin^2 x + \cos^2 x = 1.$$

PROOF Simply substitute into the left side the formulas for the sine and cosine functions that were displayed before the proposition, then simplify the result. ∎

We list several other properties of the *sine* and *cosine* functions that may be proved by similar methods. The proofs are requested of you in the Exercises.

PROPOSITION 10.15
The functions sine and cosine have the following properties:

1. $\sin(s + t) = \sin s \cos t + \cos s \sin t$
2. $\cos(s + t) = \cos s \cos t - \sin s \sin t$
3. $\cos(2s) = \cos^2 s - \sin^2 s$
4. $\sin(2s) = 2 \sin s \cos s$
5. $\sin(-s) = -\sin s$
6. $\cos(-s) = \cos s$
7. $\sin'(s) = \cos s$
8. $\cos'(s) = -\sin s$

One important task to be performed in a course on the foundations of analysis is to define the number π and establish its basic properties. In a course on Euclidean geometry, π is defined to be the ratio of the circumference of a circle to its diameter. Such a definition is not useful for our purposes (however it *is* consistent with the definition given here).

Observe that $\cos 0$ is the real part of e^{i0}, which is 1. Also there exists an $x > 0$ such that $\cos x = 0$ (why?).

$$\alpha = \inf\{x > 0 : \cos x = 0\}$$

then $\alpha > 0$ and, by the continuity of the cosine function, $\cos \alpha = 0$. We define $\pi = 2\alpha$.

Applying Proposition 10.14 to the number α yields that $\sin\alpha = \pm 1$. Since α is the *first* zero of cosine on the right half line, the cosine function must be positive on $(0, \alpha)$. But cosine is the derivative of sine. Thus the sine function is *increasing* on $(0, \alpha)$. Since $\sin 0$ is the imaginary part of e^{i0}, which is 0, we conclude that $\sin\alpha > 0$, hence that $\sin\alpha = +1$.

Now we may apply parts (3) and (4) of Proposition 10.15 with $s = \alpha$ to conclude that $\sin\pi = 0$ and $\cos\pi = -1$. A similar calculation with $s = \pi$ shows that $\sin 2\pi = 0$ and $\cos 2\pi = 1$. Next we may use parts (1) and (2) of Proposition 10.15 to calculate that $\sin(x + 2\pi) = \sin x$ and $\cos(x + 2\pi) = \cos x$ for all x. In other words, the sine and cosine functions are 2π-periodic.

The business of calculating a decimal expansion for π would take us far afield. One approach would be to utilize the already noted fact that the sine function is strictly increasing on the interval $[0, \pi/2]$, hence its inverse function

$$\mathrm{Sin}^{-1} : [0, 1] \to [0, \pi/2]$$

is well defined. Then one can determine (see Chapter 7) that

$$\left(\mathrm{Sin}^{-1}\right)'(x) = \frac{1}{\sqrt{1 - x^2}}.$$

By the Fundamental Theorem of Calculus,

$$\frac{\pi}{4} = \mathrm{Sin}^{-1}(1) = \int_0^1 \frac{1}{\sqrt{1 - x^2}}\, dx.$$

By approximating the integral by its Riemann sums, one obtains an approximation to $\pi/4$ and hence to π itself. This approach will be explored in more detail in the Exercises.

Let us for now observe that

$$\cos 2 = 1 - \frac{2^2}{2!} + \frac{2^4}{4!} - \frac{2^6}{6!} + - \cdots$$

$$= 1 - 2 + \frac{16}{24} - \frac{64}{720} + \cdots.$$

As we noted in Chapter 4, since the series defining $\cos 2$ is an alternating series with terms that strictly decrease to zero in magnitude, we may conclude that the last line is less than the sum of the first three terms:

$$\cos 2 < -1 + \frac{2}{3} < 0.$$

It follows that $\alpha = \pi/2 < 2$, hence $\pi < 4$. A similar calculation of $\cos(3/2)$ would allow us to conclude that $\pi > 3$.

10.4   Logarithms and Powers of Real Numbers

Since the exponential function $\exp(x) = e^x$ is positive and strictly increasing, it is a one-to-one function from $\mathbb{R}$ to $(0, \infty)$. Thus it has a well-defined inverse function that we call the *natural logarithm*. We write this function as $\ln x$.

PROPOSITION 10.16
The natural logarithm function has the following properties:

1. $(\ln x)' = 1/x$.

2. $\ln x$ *is strictly increasing.*

3. $\ln(1) = 0$.

4. $\ln e = 1$.

5. *The graph of the natural logarithm function is asymptotic to the negative y-axis.*

6. $\ln(s \cdot t) = \ln s + \ln t$.

7. $\ln(s/t) = \ln s - \ln t$.

PROOF   These follow immediately from corresponding properties of the exponential function.   ∎

PROPOSITION 10.17
If a and b are positive real numbers then

$$a^b = e^{b \cdot \ln a}.$$

PROOF   When b is an integer then the formula may be verified directly using Proposition 10.11. For $b = m/n$ a rational number the formula follows by our usual trick of passing to n^{th} roots. For arbitrary b we use a limiting argument as in our discussions of exponentials in Sections 3.4 and 10.3.   ∎

REMARK 10.18   We have discussed several different approaches to the exponentiation process. We proved the existence of nth roots, $n \in \mathbb{N}$, as an illustration of the completeness of the real numbers (by taking the supremum of a certain set). We treated rational exponents by composing the usual arithmetic process of taking m^{th} powers with the process of taking n^{th} roots. Then, in Sections 3.4 and 10.3, we passed to arbitrary powers by way of a limiting process.

Proposition 10.17 gives us a unified and direct way to treat all exponentials at once. This unified approach will prove (see the next proposition) to be particularly advantageous when we wish to perform calculus operations on exponential functions.   ∎

PROPOSITION 10.19

Fix $a > 0$. The function $f(x) = a^x$ has the following properties:

1. $(a^x)' = a^x \cdot \ln a$.
2. $f(0) = 1$.
3. If $0 < a < 1$ then f is decreasing and the graph of f is asymptotic to the positive x-axis.
4. If $1 < a$ then f is increasing and the graph of f is asymptotic to the negative x-axis.

PROOF   These properties follow immediately from corresponding properties of the function exp.   ∎

The logarithm function arises, among other places, in the context of probability and in the study of entropy. The reason is that the logarithm function is uniquely determined by the way that it interacts with the operation of multiplication:

THEOREM 10.20

Let $\phi(x)$ be a continuously differentiable function with domain the positive reals and that satisfies the equality

$$\phi(s \cdot t) = \phi(s) + \phi(t) \tag{$*$}$$

for all positive s and t. Then there is a constant $C > 0$ such that

$$f(x) = C \cdot \ln x$$

for all x.

PROOF   Differentiate the equation $(*)$ with respect to s to obtain

$$t \cdot \phi'(s \cdot t) = \phi'(s).$$

Now fix s and set $t = 1/s$ to conclude that

$$\phi'(s) = \phi'(1) \cdot \frac{1}{s}.$$

We take the constant C to be $\phi'(1)$ and apply Corollary 7.9 to conclude that $\phi(s) = C \cdot \ln s + D$ for some constant D. But ϕ cannot satisfy $(*)$ unless $D = 0$, so the theorem is proved.   ∎

Observe that the natural logarithm function is then the unique continuously differentiable function that satisfies the condition $(*)$ and whose derivative at 1 equals 1. That is the reason that the natural logarithm function (rather than the common logarithm, or logarithm to the base ten) is singled out as the focus of our considerations in this section.

10.5   The Gamma Function and Stirling's Formula

DEFINITION 10.4   *For $x > 0$ we define*

$$\Gamma(x) = \int_0^\infty e^{-t} t^{x-1}\, dt.$$

Notice that, by Proposition 10.12 part (6), the integrand for fixed x is majorized by the function

$$f(t) = \begin{cases} t^{x-1} & \text{if } 0 < t \le 1 \\ c_N' \cdot t^{x-N-1} & \text{if } 1 < t < \infty. \end{cases}$$

We choose N so large that $x - N - 1 < -2$. Then the function f is clearly integrable on compact intervals. By the comparison test established in Exercise 11 of Chapter 8, we conclude that the integral defining Γ converges.

PROPOSITION 10.21
For $x > 0$ we have

$$\Gamma(x + 1) = x \cdot \Gamma(x).$$

PROOF   We integrate by parts:

$$\Gamma(x + 1) = \int_0^\infty e^{-t} \cdot t^x\, dt$$

$$= \lim_{R \to +\infty} \int_0^R e^{-t} \cdot t^x\, dt$$

$$= \lim_{R \to +\infty} \left(-e^{-t} \cdot t^x \Big|_0^R + \int_0^R e^{-t} \cdot x \cdot t^{x-1}\, dt \right)$$

$$= 0 + x \cdot \Gamma(x). \qquad \blacksquare$$

COROLLARY 10.22
For $n = 1, 2, \ldots$ we have $\Gamma(n + 1) = n!$.

PROOF   An easy calculation shows that $\Gamma(1) = 1$. With induction the proposition then implies the result. $\blacksquare$

The corollary shows that the gamma function Γ is an extension of the factorial function from the positive integers to the positive real numbers. One of the exercises at the end of the chapter will ask you to verify that the gamma function is real analytic on its domain.

THEOREM 10.23   STIRLING'S FORMULA
The limit

$$\lim_{n \to \infty} \left\{ \frac{n!}{\sqrt{2\pi} e^{-n} n^{n+1/2}} \right\}$$

exists and equals 1. *In particular, the value of* $n!$ *is asymptotically equal to*

$$\frac{\sqrt{2\pi} n^{n+1/2}}{e^n}$$

as n *becomes large.*

REMARK 10.24   Stirling's formula is important in calculating limits, because without the formula it is difficult to estimate the size of $n!$ for large n. In this capacity it plays an important role in probability theory, for instance when one is examining the probable outcome of an event after a very large number of trials.

We present a particularly brief proof of Stirling's formula using the gamma function. There are a number of other proofs, some of which use complex analysis and some of which use direct estimation.   ∎

PROOF OF STIRLING'S FORMULA   Fix $x > 0$. Perform the change of variable $t = x + s\sqrt{2x}$ in the equation

$$\Gamma(x+1) = \int_0^\infty e^{-t} t^x \, dt$$

to obtain

$$\Gamma(x+1) = \sqrt{2} x^{x+1/2} e^{-x} \int_{-\sqrt{x/2}}^\infty e^{-s\sqrt{2x}} \left(1 + s\sqrt{2/x}\right)^x ds.$$

We rewrite the integrand as

$$e^{-s^2 \left(\frac{2}{s^2 \cdot (2/x)}\right) \cdot \left(s \cdot \sqrt{\frac{2}{x}}\right)} \cdot e^{x \cdot \ln\left(1 + s\sqrt{\frac{2}{x}}\right)} = e^{-s^2 \left(\frac{2}{s^2 \cdot (2/x)}\right) \cdot \left(s\sqrt{\frac{2}{x}} - \ln\left(1 + s\sqrt{\frac{2}{x}}\right)\right)}$$

$$= e^{-s^2 q\left(s\sqrt{\frac{2}{x}}\right)},$$

where q is defined by the equation

$$q(u) = \frac{2}{u^2} \cdot [u - \ln(1+u)] \ , \ u > 0.$$

By l'Hôpital's Rule, $q(u) \to 1$ as $u \to 0^+$. As $x \to +\infty$, the domain of integration $[-\sqrt{x/2}, \infty)$ expands to $(-\infty, \infty)$; the integrand tends, uniformly on compact sets of s, to e^{-s^2} (because the argument of q tends to 0). It follows (details are explored in the Exercises) that

$$\frac{\Gamma(x+1)}{\sqrt{2}x^{x+1/2}e^{-x}} \to \int_{-\infty}^{\infty} e^{-s^2}\, ds.$$

Thus our theorem is proved if we can evaluate the integral.

Set $S = \int_{-\infty}^{\infty} e^{-s^2}\, ds$. Then

$$S \cdot S = \int_{-\infty}^{\infty} e^{-x^2}\, dx \cdot \int_{-\infty}^{\infty} e^{-y^2}\, dy = \int_{-\infty}^{\infty} \int_{-\infty}^{\infty} e^{-(x^2+y^2)}\, dx\, dy.$$

We introduce polar coordinates into this two-dimensional integral:

$$S^2 = \int_{0}^{\infty} \int_{0}^{2\pi} e^{-r^2} r\, d\theta dr$$

$$= \pi \int_{0}^{\infty} e^{-r^2} 2r\, dr$$

$$= \lim_{N \to \infty} -\pi e^{-r^2} \Big|_{0}^{N}$$

$$= \pi.$$

It follows that $S = \sqrt{\pi}$ and we are done. ∎

COROLLARY 10.25
We have $\Gamma(1/2) = \sqrt{\pi}$.

PROOF Perform the change of variable $t = s^2$ in the integral defining $\Gamma(1/2)$. Then use the calculation of S in the proof of Stirling's formula. ∎

We invite the reader to justify the use of polar coordinates in the proof of Theorem 10.23.

10.6 An Introduction to Fourier Series

In this section it will be convenient for us to work on the interval $[0, 2\pi]$. We will perform arithmetic operations on this interval *modulo* 2π: for example, $3\pi/2 + 3\pi/2$ is understood to equal π because we subtract from the answer the largest multiple of 2π that it exceeds. When we refer to a function f being continuous on $[0, 2\pi]$, we require that it be right continuous at 0, left continuous at 1, and that $f(0) = f(2\pi)$.

If f is a (either real or complex-valued) Riemann integrable function on this interval and if $n \in \mathbb{Z}$, then we define

$$\hat{f}(n) = \frac{1}{2\pi} \int_0^{2\pi} f(t) e^{-int} \, dt.$$

We call $\hat{f}(n)$ the nth *Fourier coefficient* of f. The formal expression

$$Sf(x) \sim \sum_{n=-\infty}^{\infty} \hat{f}(n) e^{inx}$$

is called the *Fourier series* of the function f. In circumstances where the Fourier series converges to the function f, some of which we shall discuss below, the series provides a decomposition of f into simple component functions. This type of analysis is of importance in the theory of differential equations, in signal processing, and in scattering theory. There is a rich theory of Fourier series that is of interest in its own right.

Observe that in case f has the special form

$$f(x) = S_N(x) = \sum_{n=-N}^{N} a_n e^{inx} \qquad (*)$$

then the coefficients a_n are given by

$$a_n = \frac{1}{2\pi} \int_0^{2\pi} f(t) e^{-int} \, dt.$$

Since, in Exercise 25 of Chapter 9, we showed that functions of the form $(*)$ are dense in the continuous functions, we might hope that the coefficients $\hat{f}(n)$ of the Fourier series of f will contain important information about f.

The other theory that you know for decomposing a function into simple components is the theory of Taylor series. However, in order for a function to have a Taylor series it must be infinitely differentiable. Even then, as we have learned, the Taylor series of a function usually does not converge, and if it does converge, its limit may not be the original function. The Fourier series of f converges to f under fairly mild hypotheses on f, and thus provides a useful tool in analysis.

The first result we shall prove about Fourier series gives a growth condition on the coefficients $\hat{f}(n)$:

PROPOSITION 10.26   BESSEL'S INEQUALITY
If f^2 is integrable then

$$\sum_{n=-N}^{N} |\hat{f}_n|^2 \le \frac{1}{2\pi} \int_0^{2\pi} |f(t)|^2 \, dt$$

for every N.

PROOF Recall that $\overline{e^{ijt}} = e^{-ijt}$ and $|a|^2 = a \cdot \bar{a}$ for $a \in \mathbb{C}$. We calculate

$$\frac{1}{2\pi} \int_0^{2\pi} |f(t) - S_N(t)|^2 \, dt$$

$$= \frac{1}{2\pi} \int_0^{2\pi} \left(f(t) - \sum_{n=-N}^N \hat{f}(n) e^{int} \right) \cdot \overline{\left(f(t) - \sum_{n=-N}^N \hat{f}(n) e^{int} \right)} \, dt$$

$$= \frac{1}{2\pi} \int_0^{2\pi} |f(t)|^2 dt - \sum_{n=-N}^N \frac{1}{2\pi} \int_0^{2\pi} f(t) e^{-int} dt \cdot \overline{\hat{f}(n)}$$

$$- \sum_{n=-N}^N \frac{1}{2\pi} \int_0^{2\pi} \overline{f(t) e^{-int} dt} \cdot \hat{f}(n) + \sum_{m,n} \frac{1}{2\pi} \int_0^{2\pi} e^{imt} \cdot e^{-int} dt$$

Now each of the first two sums equals $\sum_{n=-N}^N |\hat{f}(n)|^2$. In the last sum, any summand with $m \neq n$ equals 0. (This is so because $\int_0^{2\pi} e^{ijt} \, dt = 0$ when j is a nonzero integer.) Thus our equation simplifies to

$$\frac{1}{2\pi} \int_0^{2\pi} |f(t) - S_N(t)|^2 \, dt = \frac{1}{2\pi} \int_0^{2\pi} |f(t)|^2 dt - \sum_{n=-N}^N |\hat{f}(n)|^2.$$

Since the left side is nonnegative, it follows that

$$\sum_{n=-N}^N |\hat{f}(n)|^2 \leq \frac{1}{2\pi} \int_0^{2\pi} |f(t)|^2 dt,$$

as desired. ∎

COROLLARY 10.27
If f^2 is integrable then the Fourier coefficients $\hat{f}(n)$ satisfy

$$\hat{f}(n) \to 0 \quad as \quad n \to \infty.$$

PROOF Since $\sum |\hat{f}(n)|^2 < \infty$, we know that $|\hat{f}(n)|^2 \to 0$. This implies the result. ∎

DEFINITION 10.5 *Let f be an integrable function on the interval $[0, 2\pi]$. We let $S_N(x)$ denote the Nth **partial sum of the Fourier series** of f:*

$$S_N f(x) = \sum_{n=-N}^N \hat{f}(n) e^{inx}.$$

Since the coefficients of the Fourier series, at least for a square integrable function, tend to zero, we might hope that the Fourier series will converge. Of course, the best circumstance would be that $S_N f \to f$ in some sense. We now turn our attention to addressing this problem.

PROPOSITION 10.28   THE DIRICHLET KERNEL

If f is integrable then

$$S_N f(x) = \frac{1}{2\pi} \int_0^{2\pi} D_N(x - t) f(t) \, dt,$$

where

$$D_N(t) = \frac{\sin(N + \frac{1}{2})}{\sin \frac{1}{2} t}.$$

PROOF   Observe that

$$S_N f(x) = \sum_{n=-N}^{N} \hat{f}(n) e^{inx}$$

$$= \sum_{n=-N}^{N} \frac{1}{2\pi} \int_0^{2\pi} f(t) e^{-int} dt \cdot e^{inx}$$

$$= \sum_{n=-N}^{N} \frac{1}{2\pi} \int_0^{2\pi} f(t) e^{in(x-t)} dt$$

$$= \frac{1}{2\pi} \int_0^{2\pi} f(t) \left[\sum_{n=-N}^{N} e^{in(x-t)} \right] dt.$$

Thus we are finished if we can show that the sum in [   ] equals $D_N(x - t)$.
    Rewrite the sum as

$$\sum_{n=0}^{N} \left(e^{i(x-t)} \right)^n + \sum_{n=0}^{N} \left(e^{-i(x-t)} \right)^n - 1.$$

Then each of these last two sums is the partial sum of a geometric series. Thus we use the formula from the proof of Proposition 4.7 to write the last line as

$$\frac{e^{i(x-t)(N+1)} - 1}{e^{i(x-t)} - 1} + \frac{e^{-i(x-t)(N+1)} - 1}{e^{-i(x-t)} - 1} - 1.$$

We put everything over a common denominator and obtain

$$\frac{\cos N(x-t) - \cos(N+1)(x-t)}{1 - \cos(x-t)}.$$

We write

$$N(x-t) = \left(\left(N + \frac{1}{2}\right)(x-t) - \frac{1}{2}(x-t)\right),$$

$$(N+1)(x-t) = \left(\left(N + \frac{1}{2}\right)(x-t) + \frac{1}{2}(x-t)\right),$$

$$(x-t) = \frac{1}{2}(x-t) + \frac{1}{2}(x-t)$$

and use the sum formula for the cosine function to find that the last line equals

$$\frac{2\sin\left((N+\frac{1}{2})(x-t)\right)\sin\left(\frac{1}{2}(x-t)\right)}{2\sin^2\left(\frac{1}{2}(x-t)\right)} = \frac{\sin(N+\frac{1}{2})(x-t)}{\sin\frac{1}{2}(x-t)}.$$

That is the desired conclusion. ∎

REMARK 10.29 We have presented this particular proof of the formula for D_N because it is the most natural. It is by no means the shortest. Another proof is explored in the Exercises.

Note also that, by a change of variable, the formula for S_N presented in the proposition can also be written as

$$S_N f(x) = \frac{1}{2\pi}\int_0^{2\pi} D_N(t) f(x-t)\, dt,$$

provided we adhere to the convention of doing all arithmetic modulo multiples of 2π. ∎

LEMMA 10.30
For any N it holds that

$$\frac{1}{2\pi}\int_0^{2\pi} D_N(t)\, dt = 1.$$

PROOF It would be quite difficult to prove this property of D_N from the formula that we just derived. However, if we look at the proof of the proposition we notice that

$$D_N(t) = \sum_{n=-N}^{N} e^{int}.$$

Hence

$$\frac{1}{2\pi} \int_0^{2\pi} D_N(t)\, dt = \frac{1}{2\pi} \int_0^{2\pi} \sum_{n=-N}^{N} e^{int}\, dt$$

$$= \sum_{n=-N}^{N} \frac{1}{2\pi} \int_0^{2\pi} e^{int}\, dt$$

$$= 1$$

because any power of e^{it}, except the zeroeth power, integrates to zero. This completes the proof. ∎

Next we prove that for a large class of functions the Fourier series converges back to the function at every point.

THEOREM 10.31

Let f be a function on $[0, 2\pi]$ that satisfies a Lipschitz condition: there is a constant $C > 0$ such that if $s, t \in [0, 2\pi]$ then

$$|f(s) - f(t)| \leq C \cdot |s - t|. \tag{$*$}$$

[Note that at 0 and 2π this condition is required to hold modulo 2π — see the remarks at the beginning of the section.] Then for every $x \in [0, 2\pi]$ it holds that

$$S_N f(x) \to f(x) \quad as \quad N \to \infty.$$

Indeed, the convergence is uniform in x.

PROOF   Fix $x \in [0, 2\pi]$. We calculate that

$$|S_N f(x) - f(x)| = \left| \frac{1}{2\pi} \int_0^{2\pi} f(x-t) D_N(t)\, dt - f(x) \right|$$

$$= \left| \frac{1}{2\pi} \int_0^{2\pi} f(x-t) D_N(t)\, dt \right.$$

$$\left. - \frac{1}{2\pi} \int_0^{2\pi} f(x) D_N(t)\, dt \right|,$$

where we have made use of the lemma. Now we combine the integrals to write

$$|S_N f(x) - f(x)| = \left| \frac{1}{2\pi} \int_0^{2\pi} [f(x-t) - f(x)] D_N(t) \, dt \right|$$

$$= \left| \frac{1}{2\pi} \int_0^{2\pi} \left[\frac{f(x-t) - f(x)}{\sin t/2} \right] \cdot \sin\left(\left(N + \frac{1}{2}\right)t\right) dt \right|$$

$$\leq \left| \frac{1}{2\pi} \int_0^{2\pi} \left[\frac{f(x-t) - f(x)}{\sin t/2} \cdot \cos \frac{t}{2} \right] \sin Nt \, dt \right|$$

$$+ \left| \frac{1}{2\pi} \int_0^{2\pi} \left[\frac{f(x-t) - f(x)}{\sin t/2} \cdot \sin \frac{t}{2} \right] \cos Nt \, dt \right|$$

$$\leq \left| \frac{1}{2\pi} \int_0^{2\pi} h(t) \sin Nt \, dt \right| + \left| \frac{1}{2\pi} \int_0^{2\pi} k(t) \cos Nt \, dt \right|,$$

where we have denoted the first expression in [ ] by $h_x(t) = h(t)$ and the second by $k_x(t) = k(t)$. We use our hypothesis ($*$) about f to see that

$$|h(t)| = \left| \frac{f(x-t) - f(x)}{t} \right| \cdot \left| \frac{t}{\sin(t/2)} \right| \cdot \left| \cos \frac{t}{2} \right| \leq C \cdot 2.$$

Here we have used the elementary fact that $2/\pi \leq |\sin u/u| \leq 1$. Thus h is a bounded function. It is obviously continuous, because f is, except perhaps at $t = 0$. So h is integrable — since it is bounded, it is even square integrable. An even easier discussion shows that k is square integrable. Therefore Corollary 10.27 applies and we may conclude that the Fourier coefficients of h and of k tend to zero. However, the integral involving h is nothing other than $(\hat{h}(N) - \hat{h}(-N))/(2i)$ and the integral involving k is precisely $(\hat{k}(N) + \hat{k}(-N))/2$. We conclude that these integrals tend to zero as $N \to \infty$; in other words,

$$|S_N f(x) - f(x)| \to 0 \quad \text{as} \quad N \to \infty.$$

Since the relevant estimates are independent of x, we see that the convergence is uniform. ∎

COROLLARY 10.32
If $f \in C^1([0, 2\pi])$ then $S_N f \to f$ uniformly.

PROOF A C^1 function, by the Mean Value Theorem, satisfies a Lipschitz condition. ∎

In fact, the proof of the theorem suffices to show that if f is a Riemann integrable function on $[0, 2\pi]$ and if f is differentiable at x, then $S_N f(x) \to f(x)$.

In the Exercises we shall explore other methods of summing Fourier series that allow us to realize all the continuous functions as the limits of certain Fourier expressions.

Example 10.6

Let $f(t) = t^2 - 2\pi t$, $0 \leq t \leq 2\pi$. Then $f(0) = f(2\pi) = 0$ and f is Lipschitz modulo 2π. Calculating the Fourier series of f, setting $t = 0$, and using the theorem reveals that

$$\sum_{j=1}^{\infty} \frac{1}{j^2} = \frac{\pi^2}{6}.$$

You are requested to provide the details.   ▯

Exercises

10.1 Prove Proposition 10.9.

10.2 Provide the details of the assertion preceding Proposition 10.12 to the effect that if we define, for any real $\mathbb{R}$,

$$e^r = \sup\{e^q : q \in \mathbb{Q}, q < r\},$$

then $e^x = \exp(x)$ for every real x.

10.3 Give another proof for the formula for $D_N(t)$ by completing the following outline:

(a) $D_N(t) = \sum_{n=-N}^{N} e^{int}$.

(b) $(e^{it} - 1) \cdot D_N(t) = e^{i(N+1)t} - e^{-iNt}$.

(c) Multiply both sides of the last equation by $e^{-it/2}$.

(d) Conclude that $D_N(t) = \frac{\sin(N+\frac{1}{2})t}{\sin(t/2)}$.

10.4 Assume that a power series converges at one of the endpoints of its interval of convergence. Use summation by parts to prove that the function defined by the power series is continuous on the half-open interval including that endpoint.

10.5 The function defined by a power series may extend continuously to an endpoint of the interval of convergence without the series converging at that endpoint. Give an example.

10.6 Prove Proposition 10.14 by completing the steps outlined in the text.

10.7 Let f be an infinitely differentiable function on an interval I. If $a \in I$ and there are positive constants C, R such that for every x in a neighborhood of a and every k it holds that

$$\left| f^{(k)}(x) \right| \leq C \cdot \frac{k!}{R^k},$$

then prove that the Taylor series of f about a converges to $f(x)$. (Hint: Estimate the error term.)

10.8 Let f be an infinitely differentiable function on an open interval I centered at a. Assume that the Taylor expansion of f about a converges to f at every point of I. Prove that there are constants C, R and a (possibly smaller) interval J centered at a such that for each $x \in J$ it holds that

$$\left|f^{(k)}(x)\right| \leq C \cdot \frac{k!}{R^k}.$$

10.9 Prove that the composition of two real analytic functions, when the composition makes sense, is also real analytic.

10.10 Prove that

$$\sin^2 x + \cos^2 x = 1$$

directly from the power series expansions.

10.11 Prove the equality $\left(\mathrm{Sin}^{-1}\right)' = 1/\sqrt{1 - x^2}$.

10.12 In analyzing the integral representation of $\Gamma(x + 1)$ in the proof of Stirling's formula we might have reasoned as follows: the integrand may be rewritten as

$$e^{-s\sqrt{2x}}\left(1 + s\sqrt{2/x}\right)^x = e^{-s\sqrt{2x}}\left[\left\{\left(1 + \frac{s\sqrt{2}}{\sqrt{x}}\right)^{\sqrt{x}}\right\}^{\sqrt{x}}\right].$$

As $x \to +\infty$ the expression in $\{\ \}$ tends to $e^{s\sqrt{2}}$, hence the expression in $[\ \]$ tends to $e^{s\sqrt{2x}}$. It follows that the entire integrand converges to 1. *What is wrong with this argument?*

10.13 Use one of the methods described at the end of Section 3 to calculate π to two decimal places.

10.14 Prove Proposition 10.15.

10.15 Prove Proposition 10.16.

10.16 Prove that condition $(*)$ of Theorem 10.20 implies that $\phi(1) = 0$. Assume that ϕ is differentiable at $x = 1$ but make no other hypothesis about the smoothness of ϕ. Prove that condition $(*)$ then implies that ϕ is differentiable at every $x > 0$.

10.17 Prove that if f^2 is integrable on $[0, 2\pi]$ then

$$\sum_{n=-\infty}^{\infty} |\hat{f}(n)|^2$$

is convergent.

10.18 If f is continuously differentiable on the interval $[0, 2\pi]$ and if $f'(0) = f'(2\pi)$ then prove that there is a constant $C > 0$ such that $|\hat{f}(n)| \leq C/|n|$. (Hint: Integrate by parts.)

10.19 Show that the hypothesis of Theorem 10.29 may be replaced with $f \in \mathrm{Lip}_\alpha([0, 2\pi])$, some $\alpha > 1/2$. More advanced techniques show that $\alpha > 0$ suffices. See Exercise 33.

10.20 If f is integrable on the interval $[0, 2\pi]$ and if N is a nonnegative integer then define

$$\sigma_N f(x) = \frac{1}{N + 1} \sum_{n=0}^{N} S_N(x).$$

This is called the Nth *Cesaro mean* for the Fourier series of f. Prove that

$$\sigma_N f(x) = \frac{1}{2\pi} \int_0^{2\pi} K_N(x - t) f(t) dt,$$

where

$$K_N(x - t) = \frac{1}{N + 1} \left\{ \frac{\sin \frac{N+1}{2}(x - t)}{\sin \frac{1}{2}(x - t)} \right\}^2.$$

10.21 Refer to Exercise 20 for notation. Prove that if $\delta > 0$, then $\lim_{N \to \infty} K_N(t) = 0$ with the limit being uniform for all $|t| \geq \delta$.

10.22 Refer to Exercise 20 for notation. Prove that $\frac{1}{2\pi} \int_0^{2\pi} |K_N(t)| dt = 1$.

10.23 Use the results of the preceding three exercises to prove that if f is continuous on $[0, 2\pi]$ and $f(0) = f(2\pi)$ then $\sigma_N f(x) \to f(x)$ uniformly on $[0, 2\pi]$. (Hint: Let $\epsilon > 0$. Choose $\delta > 0$ such that $|s - t| < \delta$ implies that $|f(s) - f(t)| < \epsilon$. Now divide the integral into the set where $|t| < \delta$ and the set where $|t| > \delta$ and imitate the proof of the Weierstrass Approximation Theorem.)

10.24 If $p(x) = \sum_{n=-N}^{N} a_n e^{inx}$, then calculate

$$\frac{1}{2\pi} \int_0^{2\pi} |p(x)|^2 dx$$

explicitly in terms of the a_n's.

10.25 If f is an integrable function on $[0, 2\pi]$ and $0 < r < 1$ then define

$$P_r f(x) = \frac{1}{2\pi} \int_0^{2\pi} P_r(x - t)(t) dt$$

where

$$P_r(x - t) = \frac{1 - r^2}{1 - 2r \cos(x - t) + r^2}.$$

Imitate your solution of Exercise 23 to prove that if f is continuous on $[0, 2\pi]$ and $f(0) = f(2\pi)$ then $P_r f(x) \to f(x)$, uniformly in x, as $r \to 1^-$.

10.26 Let $f(x) = \sum_{j=0}^{\infty} a_j x^j$ be defined by a power series convergent on the interval $(-r, r)$ and let Z denotes those points in the interval where f vanishes. Prove that if Z has an accumulation point in the interval, then $f \equiv 0$. (Hint: If a is the accumulation point, expand f in a power series about a. What is the first nonvanishing term in that expansion?)

10.27 Prove that if a function on an interval I has derivatives of all orders that are positive at every point of I, then f is real analytic on I.

10.28 Formulate and prove a convergence theorem for integrals that will justify the last step in the proof of Stirling's formula.

10.29 Verify that the function

$$f(x) = \begin{cases} 0 & \text{if } x = 0 \\ e^{-1/x^2} & \text{if } x \neq 0 \end{cases}$$

is infinitely differentiable on all of $\mathbb{R}$ and that $f^{(k)}(0) = 0$ for every k.

10.30 Provide the details of the proof of Proposition 10.13.

10.31 Prove that $\Gamma(x)$ is real analytic on the set $(0, \infty)$.

10.32 Complete the following outline of a proof of Ivan Niven (see [NIV]) that π is irrational:

(a) Define

$$f(x) = \frac{x^n(1 - x)^n}{n!},$$

where n is a positive integer to be selected later. For each $0 < x < 1$ we have

$$0 < f(x) < 1/n!. \qquad (*)$$

(b) For every positive integer j we have $f^{(j)}(0)$ is an integer.

(c) $f(1 - x) = f(x)$, hence $f^{(j)}(1)$ is an integer for every positive integer j.

(d) Seeking a contradiction, assume that π is rational. Then π^2 is rational. Thus we may write $\pi^2 = a/b$, where a, b are positive integers and the fraction is in lowest terms.

(e) Define

$$F(x) = b^n \left(\pi^{2n} f(x) - \pi^{2n-2} f^{(2)}(x) + \pi^{2n-4} f^{(4)}(x) - \cdots + (-1)^n f^{(2n)}(x) \right).$$

Then $F(0)$ and $F(1)$ are integers.

(f) We have

$$\frac{d}{dx} \left[F'(x) \sin(\pi x) - \pi F(x) \cos(\pi x) \right] = \pi^2 a^n f(x) \sin(\pi x).$$

(g) We have

$$\pi a^n \int_0^1 f(x) \sin(\pi x)\, dx = \left[\frac{F'(x) \sin x}{\pi} - F(x) \cos \pi x \right]_0^1 = F(1) + F(0).$$

(h) From this and $(*)$ we conclude that

$$0 < \pi a^n \int_0^1 f(x) \sin(\pi x)\, dx < \frac{\pi a^n}{n!} < 1.$$

When n is sufficiently large this contradicts the fact that $F(0) + F(1)$ is an integer.

10.33 If f is integrable on $[0, 2\pi]$ then prove that $\hat{f}(n) \to 0$ as $n \to \infty$. Use this result to prove the last assertion of Exercise 19. [First check the case that f is a trigonometric polynomial.]

11

Functions of Several Variables

11.1 Review of Linear Algebra

When we first learn linear algebra, the subject is difficult because it is not usually presented in the context of applications. Now we will see one of the most important applications of linear algebra: to provide a language in which to do analysis of several real variables. We first give a quick review of linear algebra.

For the sake of concreteness, and to keep notation as simple as possible, we only consider the vector space $\mathbb{R}^3$. Of course, everything that we do in this chapter translates easily to any other dimension. And in certain examples we will find it convenient to consider functions on $\mathbb{R}^2$ or $\mathbb{R}$.

The principal properties of a vector space are that it have an additive structure and an operation of scalar multiplication. If $\mathbf{u} = (u_1, u_2, u_3)$ and $\mathbf{v} = (v_1, v_2, v_3)$ are elements of $\mathbb{R}^3$ and $a \in \mathbb{R}$, then define the operations of addition and scalar multiplication as follows:

$$\mathbf{u} + \mathbf{v} = (u_1 + v_1, u_2 + v_2, u_3 + v_3)$$

and

$$a \cdot \mathbf{u} = (au_1, au_2, au_3).$$

Notice that the vector $\mathbf{0} = (0, 0, 0)$ is the additive identity: $\mathbf{0} + \mathbf{u} = \mathbf{u} + \mathbf{0} = \mathbf{u}$ for any element $\mathbf{u} \in \mathbb{R}^3$. Also every element $\mathbf{u} = (u_1, u_2, u_3) \in \mathbb{R}^3$ has an additive inverse $-\mathbf{u} = (-u_1, -u_2, -u_3)$ that satisfies $\mathbf{u} + (-\mathbf{u}) = \mathbf{0}$.

Example 11.1
We have

$$(3, -2, 7) + (4, 1, -9) = (7, -1, -2)$$

and

$$5 \cdot (3, -2, 7) = (15, -10, 35).$$ □

The first major idea in linear algebra is that of linear dependence:

DEFINITION 11.2 *A collection of elements* $\mathbf{u}_1, \mathbf{u}_2, \ldots, \mathbf{u}_k \in \mathbb{R}^3$ *is said to be* **linearly dependent** *if there exist constants* $a_1, a_2, \ldots, a_k$, *not all zero, such that*

$$\sum_{j=1}^{k} a_j \mathbf{u}_j = 0.$$

Example 11.3

The vectors $\mathbf{u} = (1, 3, 4), \mathbf{v} = (2, -1, -3)$, and $\mathbf{w} = (5, 1, -2)$ are linearly dependent because $1 \cdot \mathbf{u} + 2 \cdot \mathbf{v} - 1 \cdot \mathbf{w} = \mathbf{0}$.

However, the vectors $\mathbf{u}' = (1, 0, 0), \mathbf{v}' = (0, 1, 1)$, and $\mathbf{w}' = (1, 0, 1)$ are *not* linearly dependent since if there were constants a, b, c such that

$$a\mathbf{u}' + b\mathbf{v}' + c\mathbf{w}' = \mathbf{0}$$

then

$$(a + c, b, b + c) = \mathbf{0}.$$

But this means that

$$
\begin{aligned}
a + c &= 0 \\
b &= 0 \\
b + c &= 0.
\end{aligned}
$$

We conclude that a, b, c must all be equal to zero. That is not allowed in the definition of linear dependence. □

A collection of vectors that is not linearly dependent is called *linearly independent*. The vectors $\mathbf{u}', \mathbf{v}', \mathbf{w}'$ in the last example are linearly independent. Any set of three linearly independent vectors in $\mathbb{R}^3$ is called a *basis* for $\mathbb{R}^3$.

How do we recognize a basis? Notice that three vectors

$$
\begin{aligned}
\mathbf{u} &= (u_1, u_2, u_3) \\
\mathbf{v} &= (v_1, v_2, v_3) \\
\mathbf{w} &= (w_1, w_2, w_3)
\end{aligned}
$$

are linearly dependent if and only if there are numbers a, b, c, not all zero, such that

$$a\mathbf{u} + b\mathbf{v} + c\mathbf{w} = \mathbf{0}.$$

This, in turn, is true if and only if the system of equations

$$
\begin{aligned}
au_1 + bv_1 + cw_1 &= 0 \\
au_2 + bv_2 + cw_2 &= 0 \\
au_3 + bv_3 + cw_3 &= 0
\end{aligned}
$$

has a nontrivial solution. But such a system has a nontrivial solution if and only if

$$
\det \begin{pmatrix} u_1 & v_1 & w_1 \\ u_2 & v_2 & w_2 \\ u_3 & v_3 & w_3 \end{pmatrix} = 0.
$$

So a basis is a set of three vectors as above such that this determinant is *not* 0.

Bases are important because if $\mathbf{u}, \mathbf{v}, \mathbf{w}$ form a basis then every element $\mathbf{x}$ of $\mathbb{R}^3$ can be expressed in one and only one way as

$$\mathbf{x} = a\mathbf{u} + b\mathbf{v} + c\mathbf{w},$$

with a, b, c scalars. We call this a representation of $\mathbf{x}$ as a linear combination of $\mathbf{u}, \mathbf{v}, \mathbf{w}$. To see that such a representation is always possible, and is unique, let $\mathbf{x} = (x_1, x_2, x_3)$ be any element of $\mathbb{R}^3$. If $\mathbf{u}, \mathbf{v}, \mathbf{w}$ form a basis then we wish to find a, b, c such that

$$\mathbf{x} = a \cdot \mathbf{u} + b \cdot \mathbf{v} + c \cdot \mathbf{w}.$$

But, as above, this leads to the system of equations

$$
\begin{aligned}
au_1 + bv_1 + cw_1 &= x_1 \\
au_2 + bv_2 + cw_2 &= x_2 \\
au_3 + bv_3 + cw_3 &= x_3.
\end{aligned}
\qquad (*)
$$

In case $\mathbf{x} = \mathbf{0}$, then because the determinant of the matrix of coefficients of this system is nonzero, it follows that the only solution to this system is $a = 0, b = 0, c = 0$. Therefore the unique representation of $\mathbf{x}$ as a linear combination of $\mathbf{u}, \mathbf{v}, \mathbf{w}$ is $\mathbf{x} = 0 \cdot \mathbf{u} + 0 \cdot \mathbf{v} + 0 \cdot \mathbf{w}$.

In case $\mathbf{x} \neq 0$, then Cramer's Rule tells us that the unique solution of the system $(*)$ is given by

$$a = \frac{\det \begin{pmatrix} x_1 & v_1 & w_1 \\ x_2 & v_2 & w_2 \\ x_3 & v_3 & w_3 \end{pmatrix}}{\det \begin{pmatrix} u_1 & v_1 & w_1 \\ u_2 & v_2 & w_2 \\ u_3 & v_3 & w_3 \end{pmatrix}} \qquad b = \frac{\det \begin{pmatrix} u_1 & x_1 & w_1 \\ u_2 & x_2 & w_2 \\ u_3 & x_3 & w_3 \end{pmatrix}}{\det \begin{pmatrix} u_1 & v_1 & w_1 \\ u_2 & v_2 & w_2 \\ u_3 & v_3 & w_3 \end{pmatrix}}$$

$$c = \frac{\det \begin{pmatrix} u_1 & v_1 & x_1 \\ u_2 & v_2 & x_2 \\ u_3 & v_3 & x_3 \end{pmatrix}}{\det \begin{pmatrix} u_1 & v_1 & w_1 \\ u_2 & v_2 & w_2 \\ u_3 & v_3 & w_3 \end{pmatrix}}$$

Notice that the nonvanishing of the determinant in the denominator is crucial for this method to work.

In practice, we will be given a basis $\mathbf{u}, \mathbf{v}, \mathbf{w}$ for $\mathbb{R}^3$ and a vector $\mathbf{x}$ and we wish to express $\mathbf{x}$ as a linear combination of $\mathbf{u}, \mathbf{v}, \mathbf{w}$. We may do so by solving a system of linear equations as above. A more elegant way to do this is to use the concept of the inverse of a matrix.

DEFINITION 11.4 *Let*

$$M = \left(m_{pq} \right)_{\substack{p=1,\ldots,k \\ q=1,\ldots,\ell}}$$

be a $k \times \ell$ matrix (where k is the number of rows, ℓ the number of columns, and m_{pq} is the element in the pth row and q^{th} column) and

$$N = (n_{rs})_{\substack{r=1,\ldots,\ell \\ s=1,\ldots,m}}$$

*be an $\ell \times m$ matrix. Then the **product** $M \cdot N$ is defined to be the matrix*

$$T = (t_{uv})_{\substack{u=1,\ldots,k \\ q=1,\ldots,m}}$$

where

$$t_{uv} = \sum_{q=1}^{\ell} m_{uq} \cdot n_{qv}.$$

Example 11.5
Let

$$M = \begin{pmatrix} 2 & 3 & 9 \\ -1 & 4 & 0 \\ 5 & -3 & 6 \\ 4 & 4 & 1 \end{pmatrix}$$

and

$$N = \begin{pmatrix} -3 & 0 \\ 2 & 5 \\ -4 & -1 \end{pmatrix}.$$

Then $T = M \cdot N$ is well defined as a 4×2 matrix. We notice, for example, that

$$t_{11} = 2 \cdot (-3) + 3 \cdot 2 + 9 \cdot (-4) = -36$$

and

$$t_{32} = 5 \cdot 0 + (-3) \cdot 5 + 6 \cdot (-1) = -21.$$

Six other easy calculations of this kind yield that

$$M \cdot N = \begin{pmatrix} -36 & 6 \\ 11 & 20 \\ -45 & -21 \\ -8 & 19 \end{pmatrix}.$$

$\Box$

DEFINITION 11.6 *Let M be a 3×3 matrix. A matrix N is called the **inverse** of M if $M \cdot N = N \cdot M = \mathbf{I}$, where*

$$\mathbf{I} = \begin{pmatrix} 1 & 0 & 0 \\ 0 & 1 & 0 \\ 0 & 0 & 1 \end{pmatrix}.$$

*When M has an inverse then it is called **invertible**.*

It follows immediately from the definition that in order for a matrix to be invertible it must be square.

PROPOSITION 11.1
Let M be a 3×3 matrix with nonzero determinant. Then M is invertible and the elements of its inverse are given by

$$n_{ij} = \frac{(-1)^{i+j} \cdot \det M(i,j)}{\det M}.$$

Here $M(i, j)$ is the 2×2 matrix obtained by deleting the jth row and i^{th} column from M.

PROOF   This is a direct calculation that we leave to the Exercises.   ∎

DEFINITION 11.7   *If M is either a matrix or a vector, then the **transpose** of M is defined as follows: If the ij^{th} entry M is m_{ij} then the ij^{th} entry of ${}^t M$ is m_{ji}. We denote the transpose of M by ${}^t M$.*

PROPOSITION 11.2
If

$$
\begin{aligned}
\mathbf{u} &= (u_1, u_2, u_3) \\
\mathbf{v} &= (v_1, v_2, v_3) \\
\mathbf{w} &= (w_1, w_2, w_3)
\end{aligned}
$$

form a basis for $\mathbb{R}^3$ then let M be the matrix of the coefficients of these vectors and M^{-1} the inverse of M (which we know exists because the determinant of the matrix is nonzero). If $\mathbf{x} = (x_1, x_2, x_3)$ is any element of $\mathbb{R}^3$ then

$$
\mathbf{x} = \alpha \cdot \mathbf{u} + \beta \cdot \mathbf{v} + \gamma \cdot \mathbf{w}
$$

where

$$
\begin{pmatrix} \alpha \\ \beta \\ \gamma \end{pmatrix} = M^{-1} \cdot {}^t \mathbf{x}.
$$

PROOF   Let A be the vector of unknown coefficients (α, β, γ). The system of equations that we need to solve to find α, β, γ can be written in matrix notation as

$$
M \cdot {}^t A = {}^t \mathbf{x}.
$$

Applying the matrix M^{-1} to both sides of this equation gives

$$
M^{-1} \cdot M \cdot {}^t A = M^{-1} \cdot {}^t \mathbf{x}
$$

or

$$
\mathbf{I} \cdot {}^t A = M^{-1} \cdot {}^t \mathbf{x}
$$

or

$$
{}^t A = M^{-1} \cdot {}^t \mathbf{x},
$$

as desired.   ∎

The *standard basis* for $\mathbb{R}^3$ consists of the vectors

$$
\begin{array}{rcl}
\mathbf{u} & = & (1,0,0) \\
\mathbf{v} & = & (0,1,0) \\
\mathbf{w} & = & (0,0,1).
\end{array} \qquad (*)
$$

If $\mathbf{x} = (x_1, x_2, x_3)$ is any element of $\mathbb{R}^3$, then we may write

$$\mathbf{x} = x_1 \cdot \mathbf{u} + x_2 \cdot \mathbf{v} + x_3 \cdot \mathbf{w}.$$

In other words, the usual coordinates with which we locate points in three-dimensional space are the coordinates with respect to the special basis $(*)$. We write this basis as $\mathbf{e}_1, \mathbf{e}_2, \mathbf{e}_3$.

If $\mathbf{x} = (x_1, x_2, x_3)$ and $\mathbf{y} = (y_1, y_2, y_3)$ are elements of $\mathbb{R}^3$, then we define

$$\|\mathbf{x}\| = \sqrt{(x_1)^2 + (x_2)^2 + (x_3)^2}$$

and

$$\mathbf{x} \cdot \mathbf{y} = x_1 y_1 + x_2 y_2 + x_3 y_3.$$

PROPOSITION 11.3   THE SCHWARZ INEQUALITY
If $\mathbf{x}$ and $\mathbf{y}$ are elements of $\mathbb{R}^3$ then

$$|\mathbf{x} \cdot \mathbf{y}| \le \|\mathbf{x}\| \, \|\mathbf{y}\|.$$

PROOF   Write out both sides and square. If all terms are moved to the right then the right side becomes a sum of perfect squares and the inequality is obvious. Details are requested of you in an exercise.   ∎

COROLLARY 11.4
Let M be any 3×3 matrix. Then there is a constant $C > 0$ such that for any $\mathbf{x} \in \mathbb{R}^3$ we have

$$\|M({}^t\mathbf{x})\| \le C\|\mathbf{x}\|.$$

PROOF   The first entry of $M^t\mathbf{x}$ is $M_1 \cdot \mathbf{x}$, where M_1 is the first row of M. Likewise, the second entry of $M^t\mathbf{x}$ is $M_2 \cdot \mathbf{x}$ and the third entry of $M^t\mathbf{x}$ is $M_3 \cdot \mathbf{x}$. The result now follows from the Schwarz Inequality, with

$$C = \max\{\|M_1\|, \|M_2\|, \|M_3\|\}. \qquad \blacksquare$$

11.2   A New Look at the Basic Concepts of Analysis

A point of $\mathbb{R}^3$ is denoted (x, y, z) or (x_1, x_2, x_3). In the analysis of functions of one real variable, the domain of a function is typically an open interval. Since any open set in $\mathbb{R}^1$ is the disjoint union of open intervals, it is natural to work in the context of intervals. Such a simple situation does not obtain in the analysis of several variables. We will need some new notation and concepts in order to study functions in $\mathbb{R}^3$:

We measure distance between two points $\mathbf{s} = (s_1, s_2, s_3)$ and $\mathbf{t} = (t_1, t_2, t_3)$ in $\mathbb{R}^3$ by the formula

$$\|\mathbf{s} - \mathbf{t}\| = \sqrt{(s_1 - t_1)^2 + (s_2 - t_2)^2 + (s_3 - t_3)^2}.$$

Of course, this notion of distance can be justified by considerations using the Pythagorean Theorem (see the Exercises), but we treat this as a definition. The distance between two points is nonnegative and equals zero if and only if the two points are identical. Moreover, there is a triangle inequality:

$$\|\mathbf{s} - \mathbf{t}\| \leq \|\mathbf{s} - \mathbf{u}\| + \|\mathbf{u} - \mathbf{t}\|.$$

We sketch a proof of this inequality in the Exercises (by reducing it to the one-dimensional triangle inequality).

DEFINITION 11.8 *If $\mathbf{x} \in \mathbb{R}^3$ and $r > 0$ then the **open ball** with center $\mathbf{x}$ and radius r is the set*

$$B(\mathbf{x}, r) = \{\mathbf{t} \in \mathbb{R}^3 : \|\mathbf{x} - \mathbf{t}\| < r\}.$$

*The **closed ball** with center $\mathbf{x}$ and radius r is the set*

$$\overline{B}(\mathbf{x}, r) = \{\mathbf{t} \in \mathbb{R}^3 : \|\mathbf{t} - \mathbf{x}\| \leq r\}.$$

DEFINITION 11.9 *A set $U \subseteq \mathbb{R}^3$ is said to be **open** if for each $\mathbf{x} \in U$ there is an $r > 0$ such that the ball $B(\mathbf{x}, r)$ is contained in U.*

Example 11.10
Let

$$S = \{\mathbf{x} = (x_1, x_2, x_3) : 1 < \|x\| < 2\}.$$

This set is open. For if $x \in S$, let $r = \min\{\|x\| - 1, 2 - \|x\|\}$. Then $B(x, r)$ is contained in S for the following reason: if $t \in B(x, r)$ then

$$\|x\| \leq \|t - x\| + \|t\|,$$

hence

$$\|t\| \geq \|x\| - \|t - x\| > \|x\| - r \geq \|x\| - (\|x\| - 1) = 1.$$

Likewise,

$$\|t\| \leq \|x\| + \|t - x\| < \|x\| + r \leq \|x\| + (2 - \|x\|) = 2.$$

It follows that $t \in S$, hence $B(x, r) \subseteq S$. We conclude that S is open.

However, a moment's thought shows that S could not be written as a disjoint union of open balls, or open cubes, or any other regular type of open set. ☐

In this chapter we consider functions with domain a set in $\mathbb{R}^3$. This means that f may be written in the form $f(x_1, x_2, x_3)$. An example of such a function is $f(x_1, x_2, x_3) = x_1 \cdot (x_2)^4 - x_3$ or $g(x_1, x_2, x_3) = (x_3)^2 \cdot \sin(x_1 \cdot x_2 \cdot x_3)$.

DEFINITION 11.11 *Let $E \subseteq \mathbb{R}^3$ be a set and let f be a real-valued function with domain E. Fix a point $\mathbf{P}$ such that either $\mathbf{P} \in E$ or $\mathbf{P}$ is an accumulation point of E (i.e., there exist $e_j \in E$ such that $e_j \to \mathbf{P}$). We say that*

$$\lim_{x \to \mathbf{P}} f(x) = \ell,$$

with ℓ a real number, if for each $\epsilon > 0$ there is a $\delta > 0$ such that when $x \in E$ and $0 < \|x - \mathbf{P}\| < \delta$ then

$$|f(x) - \ell| < \epsilon.$$

Compare this definition with the definition in Section 6.1: the only difference is that we now measure the distance between points of the domain of f using $\| \; \|$ instead of $| \; |$.

Example 11.12
The function

$$f(x_1, x_2, x_3) = \begin{cases} \frac{x_1 x_2}{x_1^2 + x_2^2 + x_3^2} & \text{if} \quad (x_1, x_2, x_3) \neq \mathbf{0} \\ 0 & \text{if} \quad (x_1, x_2, x_3) = \mathbf{0} \end{cases}$$

has no limit as $\mathbf{x} \to \mathbf{0}$. For if we take $\mathbf{x} = (t, 0, 0)$ then we obtain the limit

$$\lim_{t \to 0} f(t, 0, 0) = 0,$$

while if we take $\mathbf{x} = (t, t, t)$ then we obtain the limit

$$\lim_{t \to 0} f(t, t, t) = \frac{1}{3}.$$

Thus for $\epsilon < \frac{1}{6} = \frac{1}{2} \cdot \frac{1}{3}$ there will exist no δ satisfying the definition of limit.

However, the function

$$g(x_1, x_2, x_3) = x_1^2 + x_2^2 + x_3^2$$

satisfies

$$\lim_{\mathbf{x} \to \mathbf{0}} g(\mathbf{x}) = 0$$

because, given $\epsilon > 0$, we take $\delta = \sqrt{\epsilon/3}$. Then $\|\mathbf{x} - \mathbf{0}\| < \delta$ implies that $|x_j - 0| < \sqrt{\epsilon/3}$ for $j = 1, 2, 3$; hence

$$|g(x_1, x_2, x_3) - 0| < \left| \left(\frac{\sqrt{\epsilon}}{\sqrt{3}} \right)^2 + \left(\frac{\sqrt{\epsilon}}{\sqrt{3}} \right)^2 + \left(\frac{\sqrt{\epsilon}}{\sqrt{3}} \right)^2 \right| = \epsilon. \qquad \square$$

Notice that, just as in the theory of one variable, the limit properties of f at a point P are independent of the *actual value* of f at P.

DEFINITION 11.13 *Let f be a function with domain E and let $P \in E$. We say that f is* **continuous** *at* $\mathbf{P}$ *if*

$$\lim_{\mathbf{x} \to \mathbf{P}} f(\mathbf{x}) = f(\mathbf{P}).$$

The limiting process respects the elementary arithmetic operations, just as in the one-variable situation explored in Chapter 6. We will treat these matters in the Exercises. Similarly, continuous functions are closed under the arithmetic operations (provided that we do not divide by zero). Next we turn to the more interesting properties of the derivative.

DEFINITION 11.14 *Let $f(\mathbf{x})$ be a function whose domain contains a ball $B(\mathbf{P}, r)$. We say that f is* **differentiable** *at* $\mathbf{P}$ *if there is a 1×3 matrix $M_\mathbf{P} = M_\mathbf{P}(f)$ such that for all $\mathbf{h} \in \mathbb{R}^3$ satisfying $\|\mathbf{h}\| < r$ it holds that*

$$f(\mathbf{P} + \mathbf{h}) = f(\mathbf{P}) + M_\mathbf{P} \cdot {}^t\mathbf{h} + \mathcal{R}_\mathbf{P}(f, \mathbf{h}),$$

where

$$\lim_{\mathbf{h} \to \mathbf{0}} \frac{\mathcal{R}_\mathbf{P}(f, \mathbf{h})}{\|\mathbf{h}\|} = 0.$$

The matrix $M_\mathbf{P} = M_\mathbf{P}(f)$ is called the **derivative** *of f at* $\mathbf{P}$.

The best way to begin to understand any new idea is to reduce it to a situation that we already understand. If f is a function of one variable that is differentiable at $\mathbf{P} \in \mathbb{R}$, then there is a number M such that

$$\lim_{h \to 0} \frac{f(\mathbf{P} + h) - f(\mathbf{P})}{h} = M.$$

We may rearrange this equality as

$$\frac{f(\mathbf{P} + h) - f(\mathbf{P})}{h} - M = \mathcal{S}_\mathbf{P},$$

where $\mathcal{S}_\mathbf{P} \to 0$ as $h \to 0$. But this may be rewritten as

$$f(\mathbf{P} + h) = f(\mathbf{P}) + M \cdot h + \mathcal{R}_\mathbf{P}(f, h), \qquad (*)$$

where

$$\lim_{h \to 0} \frac{\mathcal{R}_{\mathbf{P}}(f, h)}{h} = 0.$$

Equation $(*)$ is parallel to the equation in Definition 11.14 that defines the concept of derivative. The role of the 1×3 matrix $M_{\mathbf{P}}$ is played by the numerical constant M. *But a numerical constant is a 1×1 matrix.* Thus our equation in one variable is a special case of the equation in three variables. In one variable, the matrix representing the derivative is just the singleton consisting of the numerical derivative.

Note in passing that (in the one-variable case) the way that we now define the derivative is closely related to the Taylor expansion. The number M is the coefficient of the first-order term in that expansion, which we know from Chapter 10 to be the first derivative.

What is the significance of the matrix $M_{\mathbf{P}}$ in our definition of derivative for a function of three variables? Suppose that f is differentiable according to Definition 11.14. Let us attempt to calculate the "partial derivative" (as in calculus) with respect to x_1 of f: we have

$$f(P_1 + h, P_2, P_3) = f(\mathbf{P}) + M_{\mathbf{P}} \cdot \begin{pmatrix} h \\ 0 \\ 0 \end{pmatrix} + \mathcal{R}_{\mathbf{P}}(f, h).$$

Rearranging this equation we have

$$\frac{f(P_1 + h, P_2, P_3) - f(\mathbf{P})}{h} = (M_{\mathbf{P}})_1 + \mathcal{S}_{\mathbf{P}},$$

where $\mathcal{S}_{\mathbf{P}} \to 0$ as $h \to 0$ and $(M_{\mathbf{P}})_1$ is the first entry of the 1×3 matrix $M_{\mathbf{P}}$.

But, letting $h \to 0$ in this last equation, we see that the partial derivative with respect to x_1 of the function f exists at P and equals $(M_{\mathbf{P}})_1$. A similar calculation shows that the partial derivative with respect to x_2 of the function f exists at P and equals $(M_{\mathbf{P}})_2$; likewise, the partial derivative with respect to x_3 of the function f exists at P and equals $(M_{\mathbf{P}})_3$. We summarize with a theorem:

THEOREM 11.5

Let f be a function defined on an open ball $B(\mathbf{P}, r)$ and suppose that f is differentiable at $\mathbf{P}$ with derivative the 1×3 matrix $M_{\mathbf{P}}$. Then the first partial derivatives of f at $\mathbf{P}$ exist and they are, respectively, the entries of $M_{\mathbf{P}}$. That is,

$$(M_{\mathbf{P}})_1 = \frac{\partial}{\partial x_1} f(\mathbf{P}) \quad , \quad (M_{\mathbf{P}})_2 = \frac{\partial}{\partial x_2} f(\mathbf{P})$$

$$(M_{\mathbf{P}})_3 = \frac{\partial}{\partial x_3} f(\mathbf{P}).$$

Unfortunately, the converse of this theorem is not true: it is possible for the partial derivatives of f to exist at a single point $\mathbf{P}$ without f being differentiable at $\mathbf{P}$ in the sense of Definition 11.14. Counterexamples will be explored in the Exercises. On the other hand, the two different notions of *continuous differentiability* are the same. We formalize this statement with a proposition:

PROPOSITION 11.6
Let f be a function defined on an open ball $B(\mathbf{P}, r)$. Assume that f is differentiable on $B(\mathbf{P}, r)$ in the sense of Definition 11.14 and that the function

$$\mathbf{x} \mapsto M_{\mathbf{x}}$$

is continuous in the sense that each of the functions

$$\mathbf{x} \mapsto (M_{\mathbf{x}})_j$$

is continuous, $j = 1, 2, 3$. Then each of the partial derivatives

$$\frac{\partial}{\partial x_1} f(\mathbf{x}) \qquad \frac{\partial}{\partial x_2} f(\mathbf{x}) \qquad \frac{\partial}{\partial x_3} f(\mathbf{x})$$

exists for $\mathbf{x} \in B(\mathbf{P}, r)$ and is continuous.

Conversely, if each of the partial derivatives exists on $B(\mathbf{P}, r)$ and is continuous there, then $M_{\mathbf{x}}$ exists at each point $\mathbf{x} \in B(\mathbf{P}, r)$ and is continuous.

PROOF   This is essentially a routine check of definitions. The only place where the continuity is used is in proving the converse: that the existence and continuity of the partial derivatives implies the existence of $M_{\mathbf{x}}$. In proving the converse you should apply the one-variable Taylor expansion to the function $t \mapsto f(x + t\mathbf{h})$.   ▌

11.3   Properties of the Derivative

The arithmetic properties of the derivative — that is the sum and difference, scalar multiplication, product, and quotient rules — are straightforward and are left to the Exercises for you to consider. However, the Chain Rule takes on a different form and requires careful consideration.

In order to consider meaningful instances of the Chain Rule, we must first discuss *vector-valued* functions. That is, we consider functions with domain a subset of $\mathbb{R}^3$ and range *either* $\mathbb{R}^1$ *or* $\mathbb{R}^2$ *or* $\mathbb{R}^3$. When we consider vector-valued functions, it simplifies notation if we consider all vectors to be column vectors. This convention will be in effect for the rest of the chapter. (Thus we will no longer use the "transpose" notation.) Note in passing that the expression $\|\mathbf{x}\|$ means the same thing for a column vector as it does for a row vector — the square root of the sum of the squares of the components. Also, $f(\mathbf{x})$ means the same thing whether $\mathbf{x}$ is written as a row vector or a column vector.

Example 11.15
Define the function

$$f(x_1, x_2, x_3) = \begin{pmatrix} (x_1)^2 - x_2 \cdot x_3 \\ x_1 \cdot (x_2)^3 \end{pmatrix}$$

This is a function with domain consisting of all triples of real numbers, or $\mathbb{R}^3$, and range consisting of all pairs of real numbers, or $\mathbb{R}^2$. For example,

$$f(-1, 2, 4) = \begin{pmatrix} -7 \\ -8 \end{pmatrix}$$ ⬚

We say that a vector-valued function of three variables

$$f(\mathbf{x}) = \begin{pmatrix} f_1(\mathbf{x}) \\ \vdots \\ f_k(\mathbf{x}) \end{pmatrix}$$

(where k is 1 or 2 or 3) is differentiable at a point $\mathbf{P}$ if each of its component functions is differentiable in the sense of Section 11.2. For example, the function

$$f(x_1, x_2, x_3) = \begin{pmatrix} x_1 \cdot x_2 \\ (x_3)^2 \end{pmatrix}$$

is differentiable at all points while the function

$$g(x_1, x_2, x_3) = \begin{pmatrix} x_2 \\ |x_3| - x_1 \end{pmatrix}$$

is not differentiable at points of the form $(x_1, x_2, 0)$.

It is a good exercise in matrix algebra (which you will be asked to do at the end of the chapter) to verify that a vector-valued function f is differentiable at a point $\mathbf{P}$ if and only if there is an $n \times m$ matrix (where m is the dimension of the domain and n the dimension of the range) $M_{\mathbf{P}}(f)$ such that

$$f(\mathbf{P} + \mathbf{h}) = f(\mathbf{P}) + M_{\mathbf{P}}(f)\mathbf{h} + \mathcal{R}_{\mathbf{P}}(f, \mathbf{h});$$

here the remainder term $\mathcal{R}_{\mathbf{P}}$ is an $n \times 1$ dimensional column vector satisfying

$$\frac{\|\mathcal{R}_{\mathbf{P}}(f, \mathbf{h})\|}{\|h\|} \to 0$$

as $h \to 0$. One nice consequence of this formula is that, by what we learned in the last section about partial derivatives, the entry in the ith row and j^{th} column of the matrix M is $\partial f_i / \partial x_j$.

Of course, the Chain Rule provides a method for differentiating compositions of functions. What we will discover in this section is that the device of thinking of the derivative as a matrix occurring in an expansion of f about a point a makes the Chain Rule a very natural and easy result to derive. It will also prove to be a useful way of keeping track of information.

THEOREM 11.7

Let f and g each be functions of three real variables and taking values in $\mathbb{R}^3$. Suppose that the range of g is contained in the domain of f, so that $f \circ g$ makes sense. If g is differentiable at a point $\mathbf{P}$ in its domain and f is differentiable at $g(\mathbf{P})$ then $f \circ g$ is differentiable at $\mathbf{P}$ and its derivative is $M_{g(\mathbf{P})}(f) \cdot M_{\mathbf{P}}(g)$. We use the symbol $\cdot$ here to denote matrix multiplication.

PROOF By the hypothesis about the differentiability of g,

$$
\begin{aligned}
(f \circ g)(\mathbf{P} + \mathbf{h}) &= f(g(\mathbf{P} + \mathbf{h})) \\
&= f\left(g(\mathbf{P}) + M_{\mathbf{P}}(g)\mathbf{h} + \mathcal{R}_{\mathbf{P}}(g, \mathbf{h})\right) \\
&= f\left(g(\mathbf{P}) + \mathbf{k}\right), \qquad\qquad\qquad (*)
\end{aligned}
$$

where

$$
\mathbf{k} = M_{\mathbf{P}}(g)\mathbf{h} + \mathcal{R}_{\mathbf{P}}(g, \mathbf{h}).
$$

But then the differentiability of f at $g(\mathbf{P})$ implies that $(*)$ equals

$$
f(g(\mathbf{P})) + M_{g(\mathbf{P})}(f)\mathbf{k} + \mathcal{R}_{g(\mathbf{P})}(f, \mathbf{k}).
$$

Now let us substitute in the value of $\mathbf{k}$. We find that

$$
\begin{aligned}
(f \circ g)(\mathbf{P} + \mathbf{h}) &= f(g(\mathbf{P})) + M_{g(\mathbf{P})}(f)[M_{\mathbf{P}}(g)\mathbf{h} + \mathcal{R}_{\mathbf{P}}(g, \mathbf{h})] \\
&\quad + \mathcal{R}_{g(\mathbf{P})}(f, M_{\mathbf{P}}(g)\mathbf{h} + \mathcal{R}_{\mathbf{P}}(g, \mathbf{h})) \\
&= f(g(\mathbf{P})) + M_{g(\mathbf{P})}(f)M_{\mathbf{P}}(g)\mathbf{h} \\
&\quad + \left\{ M_{g(\mathbf{P})}(f)\mathcal{R}_{\mathbf{P}}(g, \mathbf{h}) \right. \\
&\quad \left. + \mathcal{R}_{g(\mathbf{P})}(f, M_{\mathbf{P}}(g)\mathbf{h} + \mathcal{R}_{\mathbf{P}}(g, \mathbf{h})) \right\} \\
&\equiv f(g(\mathbf{P})) + M_{g(\mathbf{P})}(f)M_{\mathbf{P}}(g)\mathbf{h} \\
&\quad + \mathcal{Q}_{\mathbf{P}}(f \circ g, \mathbf{h}),
\end{aligned}
$$

where the last equality defines $\mathcal{Q}$. The term $\mathcal{Q}$ should be thought of as a remainder term. Since

$$
\frac{\|\mathcal{R}_{\mathbf{P}}(g, \mathbf{h})\|}{\|\mathbf{h}\|} \to 0
$$

as $\mathbf{h} \to 0$ it follows that

$$\frac{M_{g(\mathbf{P})}(f)\mathcal{R}_{\mathbf{P}}(g, \mathbf{h})}{\|\mathbf{h}\|} \to 0.$$

(Details of this assertion are requested of you in the Exercises.) Similarly,

$$\frac{\mathcal{R}_{g(\mathbf{P})}(f, M_{\mathbf{P}}(g)\mathbf{h} + \mathcal{R}_{\mathbf{P}}(g, \mathbf{h}))}{\|\mathbf{h}\|} \to 0$$

as $\mathbf{h} \to 0$.

It follows that $f \circ g$ is differentiable at $\mathbf{P}$ and that the derivative equals $M_{g(\mathbf{P})}(f)M_{\mathbf{P}}(g)$, the product of the derivatives of f and g. ∎

REMARK 11.8 Notice that, by our hypotheses, $M_{\mathbf{P}}(g)$ is a 3×3 matrix and so is $M_{g(\mathbf{P})}(f)$. Thus their product makes sense.

In general, if g is a function from a subset of $\mathbb{R}^m$ to $\mathbb{R}^n$ then, if we want $f \circ g$ to make sense, f must be a function from a subset of $\mathbb{R}^n$ to some $\mathbb{R}^k$. In other words, the dimension of the range of g had better match the dimension of the domain of f. Then the derivative of g at some point $\mathbf{P}$ will be an $n \times m$ matrix and the derivative of f at $g(\mathbf{P})$ will be a $k \times n$ matrix. Then the matrix multiplication $M_{g(\mathbf{P})}(f)M_{\mathbf{P}}(g)$ will make sense. ∎

COROLLARY 11.9 THE CHAIN RULE IN COORDINATES
Let $f : \mathbb{R}^3 \to \mathbb{R}^3$ and $g : \mathbb{R}^3 \to \mathbb{R}^3$ be vector-valued functions and assume that $h = f \circ g$ makes sense. If g is differentiable at a point $\mathbf{P}$ of its domain and f is differentiable at $g(\mathbf{P})$ then for each i and j we have

$$\frac{\partial h_i}{\partial x_j}(\mathbf{P}) = \sum_{m=1}^{3} \frac{\partial f_i}{\partial s_m}(g(\mathbf{P})) \cdot \frac{\partial g_m}{\partial x_j}(\mathbf{P}).$$

PROOF The function $\partial h_i / x_j$ is the entry of $M_{\mathbf{P}}(h)$ in the ith row and j^{th} column. However, $M_{\mathbf{P}}(h)$ is the product of $M_{g(\mathbf{P})}(f)$ with $M_{\mathbf{P}}(g)$. The entry in the ith row and j^{th} column of that product is

$$\sum_{m=1}^{3} \frac{\partial f_i}{\partial s_m}(g(\mathbf{P})) \cdot \frac{\partial g_m}{\partial x_j}(\mathbf{P}). \qquad ∎$$

We conclude this section by deriving a Taylor expansion for scalar-valued functions of three real variables: this expansion for functions of several variables is derived in an interesting way from the expansion for functions of one variable. We say that a function f of several real variables is k times continuously differentiable if all partial derivatives of orders up to and including k exist and are continuous on the domain of f.

THEOREM 11.10 TAYLOR'S EXPANSION

For k a nonnegative integer let f be a $k + 1$ times continuously differentiable scalar-valued function on a neighborhood of a closed ball $\overline{B}(\mathbf{P}, r) \subseteq \mathbb{R}^3$. Then, for $x \in B(\mathbf{P}, r)$,

$$f(x) = \sum_{0 \le j_1 + j_2 + j_3 \le k} \frac{\partial^{j_1 + j_2 + j_3} f}{\partial x_1^{j_1} \partial x_2^{j_2} \partial x_3^{j_3}}(\mathbf{P}) \cdot \frac{(x_1 - P_1)^{j_1}(x_2 - P_2)^{j_2}(x_3 - P_3)^{j_3}}{(j_1)!(j_2)!(j_1)!}$$
$$+ \mathcal{R}_{k,\mathbf{P}}(x),$$

where

$$|\mathcal{R}_{k,\mathbf{P}}(x)| \le C_0 \cdot \frac{\|\mathbf{x} - \mathbf{P}\|^{k+1}}{(k+1)!},$$

and

$$C_0 = \sup_{j_1 + j_2 + j_3 = k+1} \; \sup_{t \in \bar{B}(\mathbf{P}, r)} \left| \frac{\partial^{j_1 + j_2 + j_3} f}{\partial x_1^{j_1} \partial x_2^{j_2} \partial x_3^{j_3}}(t) \right|.$$

PROOF   With $\mathbf{P}$ and $\mathbf{x}$ fixed, define

$$\mathcal{F}(t) = f(\mathbf{P} + t(\mathbf{x} - \mathbf{P})) \quad 0 \le t < \frac{r}{\|\mathbf{x} - \mathbf{P}\|}.$$

We apply the one-dimensional Taylor theorem to the function $\mathcal{F}$, expanded about the point 0:

$$\mathcal{F}(t) = \sum_{\ell=0}^{k} \mathcal{F}^{(\ell)}(0) \frac{t^\ell}{\ell!} + R_{k,0}(\mathcal{F}, t).$$

Now the Chain Rule shows that

$$\mathcal{F}^{(\ell)}(0) =$$

$$\sum_{j_1 + j_2 + j_3 = \ell} \frac{\partial^{j_1 + j_2 + j_3} f}{\partial x_1^{j_1} \partial x_2^{j_2} \partial x_3^{j_3}}(\mathbf{P}) \cdot \frac{\ell!}{(j_1)!(j_2)!(j_3)!} \cdot (x_1 - P_1)^{j_1}(x_2 - P_2)^{j_2}(x_3 - P_3)^{j_3}.$$

Substituting this last equation, for each ℓ, into the formula for $\mathcal{F}(t)$ and setting $t = 1$ (recall that $r/\|\mathbf{x} - \mathbf{P}\| > 1$) yields the desired expression for $f(x)$. It remains to estimate the remainder term.

The one-variable Taylor theorem tells us that, for $t > 0$,

$$|R_{k,0}(\mathcal{F}, t)| = \left| \int_0^t \mathcal{F}^{(k+1)}(s) \frac{(t-s)^k}{k!} \, ds \right|$$

$$\leq \int_0^t C_0 \cdot \|\mathbf{x} - \mathbf{P}\|^{k+1} \cdot \left| \frac{(t-s)^k}{k!} \right| \, ds$$

$$= C_0 \cdot \frac{\|\mathbf{x} - \mathbf{P}\|^{k+1}}{(k+1)!}.$$

Here we have, of course, used the Chain Rule to pass from derivatives of $\mathcal{F}$ to derivatives of f. This is the desired result. ∎

11.4   The Inverse and Implicit Function Theorems

It is easy to tell whether a continuous function of one real variable is invertible. If the function is strictly monotone increasing or strictly monotone decreasing on an interval then the restriction of the function to that interval is invertible. The converse is true as well. It is more difficult to tell whether a function of several variables, when restricted to a neighborhood of a point, is invertible. The reason, of course, is that such a function will in general have different monotonicity behavior in different directions.

However, if we look at the one-variable situation in a new way it can be used to give us an idea for analyzing functions of several variables. Suppose that f is continuously differentiable on an open interval I and that $P \in I$. If $f'(P) > 0$ then the continuity of f' tells us that, for x near P, $f'(x) > 0$. Thus f is strictly monotone increasing on some (possibly smaller) open interval J centered at P. Such a function, when restricted to J, is an invertible function. The same analysis applies when $f'(P) < 0$.

Now the hypothesis that $f'(P) > 0$ or $f'(P) < 0$ has an important geometric interpretation — the positivity of $f'(P)$ means that the tangent line to the graph of f at P has positive slope, hence that the tangent line is the graph of an invertible function; likewise, the negativity of $f'(P)$ means that the tangent line to the graph of f at P has negative slope, hence that the tangent line is the graph of an invertible function (draw a sketch). Since the tangent line is a very close approximation at P to the graph of f, our geometric intuition suggests that the local invertibility of f is closely linked to the invertibility of the function describing the tangent line. This guess is in fact borne out in the discussion in the last paragraph.

We would like to carry out an analysis of this kind for a function f from a subset of $\mathbb{R}^3$ into $\mathbb{R}^3$. If P is in the domain of f and if a certain derivative of f at P (to be discussed below) does not vanish, then we would like to conclude that there is a neighborhood U of P such that the restriction of f to U is invertible. That is the content of the Inverse Function Theorem.

Before we formulate and prove this important theorem, we first discuss the kind of derivative of f at P that we shall need to examine.

DEFINITION 11.16   *Let f be a differentiable function from an open subset U of $\mathbb{R}^3$ into $\mathbb{R}^3$. The **Jacobian matrix** of f at a point $\mathbf{P} \in U$ is the matrix*

$$Jf(\mathbf{P}) = \begin{pmatrix} \frac{\partial f_1}{\partial x_1}(\mathbf{P}) & \frac{\partial f_1}{\partial x_2}(\mathbf{P}) & \frac{\partial f_1}{\partial x_3}(\mathbf{P}) \\ \frac{\partial f_2}{\partial x_1}(\mathbf{P}) & \frac{\partial f_2}{\partial x_2}(\mathbf{P}) & \frac{\partial f_2}{\partial x_3}(\mathbf{P}) \\ \frac{\partial f_3}{\partial x_1}(\mathbf{P}) & \frac{\partial f_3}{\partial x_2}(\mathbf{P}) & \frac{\partial f_3}{\partial x_3}(\mathbf{P}) \end{pmatrix}.$$

Notice that if we were to expand the function f in a Taylor series about $\mathbf{P}$ (this would be in fact a triple of expansions, since $f = (f_1, f_2, f_3)$) then the expansion would be

$$f(\mathbf{P} + \mathbf{h}) = f(\mathbf{P}) + Jf(P)\mathbf{h} + \dots.$$

Thus the Jacobian matrix is a natural object to study. It is the same object that we called $M_{\mathbf{P}}(f)$ in earlier sections. Moreover, we see that the expression $f(\mathbf{P} + \mathbf{h}) - f(\mathbf{P})$ is well approximated by the expression $Jf(\mathbf{P})\mathbf{h}$. Thus, in analogy with one-variable analysis, we might expect that the invertibility of the matrix $Jf(\mathbf{P})$ would imply the existence of a neighborhood of $\mathbf{P}$ on which the function f is invertible. This is indeed the case:

THEOREM 11.11   THE INVERSE FUNCTION THEOREM
Let f be a continuously differentiable function from an open set $U \subseteq \mathbb{R}^3$ into $\mathbb{R}^3$. Suppose that $\mathbf{P} \in U$ and that the matrix $Jf(\mathbf{P})$ is invertible. Then there is a neighborhood V of $\mathbf{P}$ such that the restriction of f to V is invertible.

PROOF   The proof of the theorem as stated is rather difficult. Therefore we shall content ourselves with the proof of a special case: we shall make the additional hypothesis that the function f is twice continuously differentiable in a neighborhood of $\mathbf{P}$.

Choose $s > 0$ such that $\overline{B}(\mathbf{P}, s) \subseteq U$ and so that $\det Jf(x) \neq 0$ for all $x \in \bar{B}(P, s)$. Thus the Jacobian matrix $Jf(x)$ is invertible for all $x \in \bar{B}(P, s)$. With the extra hypothesis, Taylor's Theorem tells us that there is a constant C such that if $\|\mathbf{Q} - \mathbf{P}\| < s/2$ then

$$f(\mathbf{Q} + \mathbf{h}) - f(\mathbf{Q}) = Jf(\mathbf{Q})\mathbf{h} + \mathcal{R}_{1,\mathbf{Q}}(f, \mathbf{Q} + \mathbf{h}), \qquad (*)$$

where

$$|\mathcal{R}_{1,\mathbf{Q}}(\mathbf{Q} + \mathbf{h})| \leq C \cdot \frac{\|\mathbf{h}\|^2}{(2)!},$$

and

$$C = \sup_{j_1+j_2+j_3=k+1} \quad \sup_{x\in \bar{B}(\mathbf{Q},s/2)} \left| \frac{\partial^{j_1+j_2+j_3} f}{\partial x_1^{j_1} \partial x_2^{j_2} \partial x_3^{j_3}}(x) \right|.$$

However, all the derivatives in the sum specifying C are, by hypothesis, continuous functions. Since all the balls $B(\mathbf{Q}, s/2)$ are contained in the compact subset $\bar{B}(P, s)$ of U, it follows that we may choose C to be a finite number *independent of* $\mathbf{Q}$.

Now the matrix $Jf(\mathbf{Q})^{-1}$ exists by hypothesis. The coefficients of this matrix will be continuous functions of $\mathbf{Q}$ because those of Jf are. Thus these coefficients will be bounded above on $\bar{B}(\mathbf{P}, s)$. By Corollary 11.4, there is a constant $K > 0$ *independent of* $\mathbf{Q}$ such that for every $\mathbf{k} \in \mathbb{R}^3$ we have

$$\|Jf(\mathbf{Q})^{-1}\mathbf{k}\| \le K\|\mathbf{k}\|.$$

Taking $\mathbf{k} = Jf(\mathbf{Q})\mathbf{h}$ yields

$$\|\mathbf{h}\| \le K\|Jf(\mathbf{Q})\mathbf{h}\|. \tag{$**$}$$

Now set

$$r = \min\{s/2, 1/(KC)\}.$$

Line $(*)$ tells us that, for $\mathbf{Q} \in B(\mathbf{P}, r)$ and $\|\mathbf{h}\| < r$,

$$\|f(\mathbf{Q} + \mathbf{h}) - f(\mathbf{Q})\| \ge \|Jf(Q)\mathbf{h}\| - \|\mathcal{R}_{1,\mathbf{Q}}(\mathbf{Q} + \mathbf{h})\|.$$

But estimate $(**)$, together with our estimate from above on the error term $\mathcal{R}$, yields that the right side of this equation is

$$\ge \frac{\|\mathbf{h}\|}{K} - \frac{C}{2}\|\mathbf{h}\|^2.$$

The choice of r tells us that $\|\mathbf{h}\| \le 1/(KC)$, hence the last line majorizes $(K/2)\|\mathbf{h}\|$.

But this tells us that for any $\mathbf{Q} \in B(\mathbf{P}, r)$ and any $\mathbf{h}$ satisfying $\|\mathbf{h}\| < r$ it holds that $f(\mathbf{Q} + \mathbf{h}) \ne f(\mathbf{Q})$. In particular, the function f is one-to-one when restricted to the ball $B(\mathbf{P}, r/2)$. Thus $f|_{B(P,s/2)}$ is invertible. ∎

In fact, the estimate

$$\|f(\mathbf{Q} + \mathbf{h}) - f(\mathbf{Q})\| \ge \frac{K}{2}\|\mathbf{h}\|$$

that we derived easily implies that the image of every $B(\mathbf{Q}, s)$ contains an open ball $B(f(\mathbf{Q}), s')$, some $s' > 0$. This means that f is an *open mapping*. You will be asked in the Exercises to provide details of this assertion. A consequence is that f^{-1} is continuous.

With some additional effort it can be shown that f^{-1} is continuously differentiable in a neighborhood of $f(\mathbf{P})$. However, the details of this matter are beyond the scope of this book. We refer the interested reader to [RUD].

Next we turn to the Implicit Function Theorem. This result addresses the question of when we can solve an equation

$$f(x_1, x_2, x_3) = 0$$

for one of the variables in terms of the other two. It is illustrative to first consider a simple example. Look at the equation

$$f(x_1, x_2) = (x_1)^2 + (x_2)^2 = 1.$$

We may restrict attention to $-1 \leq x_1 \leq 1$, $-1 \leq x_2 \leq 1$. As a glance at the graph shows, we can solve this equation for x_2, uniquely in terms of x_1, in a neighborhood of any point *except* for the points $(\pm 1, 0)$. At these two exceptional points, it is impossible to avoid the ambiguity in the square root process, even by restricting to a very small neighborhood. At other points, we may write

$$t_2 = \sqrt{1 - (t_1)^2}$$

for points (t_1, t_2) near (x_1, x_2) when $x_2 > 0$ and

$$t_2 = -\sqrt{1 - (t_1)^2}$$

for points (t_1, t_2) near (x_1, x_2) when $x_2 < 0$.

What distinguishes the two exceptional points from the others is that the tangent line to the locus (a circle) is vertical at each of these points. Another way of saying this is that

$$\frac{\partial f}{x_2} = 0$$

at these points. These preliminary considerations motivate the following theorem.

THEOREM 11.12   THE IMPLICIT FUNCTION THEOREM

Let f be a function of three real variables, taking scalar values, whose domain contains a neighborhood of a point $\mathbf{P}$. Assume that f is continuously differentiable and that $f(\mathbf{P}) = 0$. If $(\partial f / \partial x_3)(\mathbf{P}) \neq 0$ then there are numbers $\delta > 0, \eta > 0$ such that if $|x_1 - P_1| < \delta$ and $|x_2 - P_2| < \delta$ then there is a unique x_3 with $|x_3 - P_3| < \eta$ and

$$f(x_1, x_2, x_3) = 0. \tag{$*$}$$

In other words, in a neighborhood of $\mathbf{P}$, the equation $(*)$ uniquely determines x_3 in terms of x_1 and x_2.

PROOF   We consider the function

$$T : (x_1, x_2, x_3) \longmapsto \left(x_1, x_2, f(x_1, x_2, x_3)\right).$$

The Jacobian matrix of T at **P** is

$$\begin{pmatrix} 1 & 0 & 0 \\ 0 & 1 & 0 \\ \frac{\partial f}{\partial x_1}(\mathbf{P}) & \frac{\partial f}{\partial x_2}(\mathbf{P}) & \frac{\partial f}{\partial x_3}(\mathbf{P}) \end{pmatrix}.$$

Of course, the determinant of this matrix is $\partial f / \partial x_3(\mathbf{P})$, which we hypothesized to be nonzero. Thus the Inverse Function Theorem applies to T. We conclude that T is invertible in a neighborhood of **P**. That is, there is a number $\eta > 0$ and a neighborhood W of the point $(P_1, P_2, 0)$ such that

$$T : B(\mathbf{P}, \eta) \longmapsto W$$

is a one-to-one, onto, continuously differentiable function that is invertible. Select $\delta > 0$ such that if $|x_1 - P_1| < \delta$ and $|x_2 - P_2| < \delta$ then the point $(x_1, x_2, 0) \in W$. Such a point $(x_1, x_2, 0)$ then has a unique inverse image under T that lies in $B(\mathbf{P}, \eta)$. But this just says that there is a unique x_3 such that $f(x_1, x_2, x_3) = 0$. We have established the existence of δ and η as required, hence the proof is complete. ∎

Exercises

11.1 Prove that any set of vectors in $\mathbb{R}^3$ that is linearly independent cannot have more than three elements.

11.2 Prove Proposition 11.1.

11.3 Prove Proposition 11.3.

11.4 Fix elements $\mathbf{s}, \mathbf{t}, \mathbf{u} \in \mathbb{R}^3$. First assume that these three points are colinear. By reduction to the one-dimensional case, prove the triangle inequality

$$\|\mathbf{s} - \mathbf{t}\| \leq \|\mathbf{s} - \mathbf{u}\| + \|\mathbf{u} - \mathbf{t}\|.$$

Now establish the general case of the triangle inequality by comparison with the colinear case.

11.5 Give another proof of the triangle inequality by squaring both sides and invoking the Schwarz inequality.

11.6 If $\mathbf{s}, \mathbf{t} \in \mathbb{R}^3$ then prove that

$$\|\mathbf{s} + \mathbf{t}\| \geq \|\mathbf{s}\| - \|\mathbf{t}\|.$$

11.7 Formulate and prove the elementary properties of limits for functions of three variables (refer to Chapter 6 for the one-variable analogues).

11.8 Formulate and prove the elementary properties (regarding addition, scalar multiplication, etc.) of continuous functions of three variables (refer to Chapter 6 for the one-variable analogues).

11.9 Prove that the Implicit Function Theorem implies the Inverse Function Theorem.

11.10 Give an example of a function f defined in a neighborhood of the origin in $\mathbb{R}^3$ for which all partial derivatives exist at 0 but f is not differentiable at 0. (Hint: f need not even be continuous at 0.)

11.11 Prove Proposition 11.6.

11.12 Prove that a vector-valued function f is differentiable at a point $\mathbf{P}$ if and only if it can be written as

$$f(\mathbf{P} + \mathbf{h}) = f(\mathbf{P}) + M_{\mathbf{P}}(f)\mathbf{h} + \mathcal{R}_{\mathbf{P}}(f, \mathbf{h})$$

as discussed in the text prior to Theorem 11.7.

11.13 Provide the details for the assertion about what the Chain Rule shows in the proof of Taylor's Expansion.

11.14 Prove that a function satisfying the hypotheses of the Inverse Function Theorem is an open mapping in a neighborhood of the point $\mathbf{P}$.

11.15 Prove that the Implicit Function Theorem is still true if the equation $f(x_1, x_2, x_3) = 0$ is replaced by $f(x_1, x_2, x_3) = c$. (Hint: Do *not* repeat the proof of the Implicit Function Theorem.)

11.16 Let $f(x_1, x_2) = ((x_1)^3 - x_1 \cdot x_2, \sin(x_1 \cdot x_2))$ and $g(x_1, x_2, x_3) = (\ln(x_1 + x_3), \cos x_2)$. Calculate all the first partial derivatives of $f \circ g$.

11.17 Give an example of an infinitely differentiable function with domain $\mathbb{R}^2$ such that $\{(x_1, x_2) : f(x_1, x_2) = 0\} = \{(x_1, x_2) : |x_1|^2 + |x_2|^2 \le 1\}$.

11.18 Formulate a definition of second derivative parallel to the definition of first derivative given in Section 11.2. Your definition should involve a matrix. What does this matrix tell us about the second partial derivatives of the function?

11.19 Formulate and prove a product rule for derivatives of functions of three variables.

11.20 Formulate and prove a sum and difference rule for derivatives of functions of three variables.

11.21 Formulate and prove a quotient rule for derivatives of functions of three variables.

11.22 If f and g are vector-valued functions both taking values in $\mathbb{R}^3$ and both having the same domain, then we can define the dot product function $h(\mathbf{x}) = f(\mathbf{x}) \cdot g(\mathbf{x})$. Formulate and prove a product rule for this type of product.

11.23 Formulate a notion of "bounded variation" for functions of two real variables. Explain why your definition is a reasonable generalization of the notion for one real variable. (This matter was originally studied by Tonelli.)

11.24 Formulate a notion of uniform convergence for functions of three real variables. Prove that the uniform limit of a sequence of continuous functions is continuous.

11.25 Formulate a notion of "compact set" for subsets of $\mathbb{R}^3$. Prove that the continuous image, under a vector-valued function, of a compact set is compact.

11.26 Refer to Exercise 25. Prove that if f is a continuous scalar-valued function on a compact set then f assumes both a maximum value and a minimum value.

11.27 Prove that if a function with domain an open subset of $\mathbb{R}^3$ is differentiable at a point P then it is continuous at P.

11.28 Justify our notion of distance in $\mathbb{R}^3$ using Pythagorean theorem considerations.

11.29 Verify the last two assertions in the proof of Theorem 11.7.

11.30 Let f be a function defined on a ball $B(\mathbf{P}, r)$. Let $\mathbf{u} = (u_1, u_2, u_3)$ be a vector of unit length. If f is differentiable at $\mathbf{P}$ then give a definition of the directional derivative $D_{\mathbf{u}} f(\mathbf{P})$ of f in the direction $\mathbf{u}$ at P in terms of $M_{\mathbf{P}}$.

11.31 If f is differentiable on a ball $B(\mathbf{P}, r)$ and if M_x is the zero matrix for every $x \in B(\mathbf{P}, r)$, then prove that f is constant on $B(\mathbf{P}, r)$.

11.32 Refer to Exercise 30 for notation. For which collections of vectors $\mathbf{u}_1, \mathbf{u}_2, \mathbf{u}_3$ is it true that if $D_{\mathbf{u}_j} f(x) = 0$ for all $x \in B(\mathbf{P}, r)$ and all $j = 1, 2, 3$, then f is identically constant?

11.33 There is no mean value theorem as such in the theory of functions of several real variables. For example, if $\gamma : [0, 1] \to \mathbb{R}^3$ is a differentiable function on $(0, 1)$, continuous on $[0, 1]$, then it is not the case that there is a point $\xi \in (0, 1)$ such that $\dot{\gamma}(\xi) = \gamma(1) - \gamma(0)$. Provide a counterexample to substantiate this claim.

However there is a serviceable substitute for the mean value theorem: if we assume that γ is continuously differentiable on an open interval that contains $[a, b]$ and if $M = \max_{t \in [a,b]} |\dot{\gamma}(t)|$, then

$$|\gamma(b) - \gamma(a)| \leq M \cdot |b - a|.$$

Prove this statement.

11.34 Let f be a continuously differentiable function with domain the unit ball in $\mathbb{R}^3$ and range $\mathbb{R}$. Let P, Q be points of the ball. Using Exercise 33 for inspiration, formulate and prove a sort of "mean value theorem" for f that estimates $|f(P) - f(Q)|$ in terms of the gradient of f.

12

Advanced Topics

12.1 Metric Spaces

As you studied Chapter 11, and did the Exercises developing the basic properties of functions of several variables, you should have noticed that many of the proofs were identical to those in Chapter 6. The arguments generally involved clever use of the triangle inequality. For functions of one variable, the inequality was for $|\ |$. For functions of several variables, the inequality was for $\|\ \|$.

This section formalizes a general context in which we may do analysis any time we have a reasonable notion of calculating distance. Such a structure will be called a metric:

DEFINITION 12.1   *A **metric space** is a pair (X, ρ), where X is a set and*

$$\rho : X \times X \to \{t \in \mathbb{R} : t \geq 0\}$$

is a function satisfying

1. $\forall x, y \in X, \rho(x, y) = \rho(y, x)$;
2. $\rho(x, y) = 0$ *if and only if* $x = y$;
3. $\forall x, y, z \in X, \rho(x, y) \leq \rho(x, z) + \rho(z, y)$.

*The function ρ is called a **metric** on X.*

Example 12.2
The pair $(\mathbb{R}, \rho)$, where $\rho(x, y) = |x - y|$, is a metric space. Each of the properties required of a metric is, in this case, a restatement of familiar facts from the analysis of one dimension.

The pair $(\mathbb{R}^3, \rho)$, where $\rho(x, y) = \|x - y\|$, is a metric space. Each of the properties required of a metric is in this case a restatement of familiar facts from the analysis of three dimensions. ▯

Example 12.2 presented familiar metrics on two familiar spaces. Now we look at some new ones.

Example 12.3

The pair $(\mathbb{R}^2, \rho)$, where $\rho(x, y) = \max\{|x_1 - y_1|, |x_2 - y_2|\}$, is a metric space. Only the triangle inequality is not trivial to verify, but that reduces to the triangle inequality of one variable.

The pair $(\mathbb{R}, \mu)$, where $\mu(x, y) = 1$ if $x \neq y$ and 0 otherwise, is a metric space. Checking the triangle inequality reduces to seeing that if $x \neq y$ then either $x \neq z$ or $y \neq z$. ▯

Example 12.4

Let X denote the space of continuous functions on the interval $[0, 1]$. If $f, g \in X$ then let $\rho(f, g) = \sup_{t \in [0,1]} |f(t) - g(t)|$. Then the pair (X, ρ) is a metric space. The first two properties of a metric are obvious and the triangle inequality reduces to the triangle inequality for real numbers.

This example is a dramatic new departure from the analysis we have done in the previous eleven chapters. For X is a very large space — infinite dimensional in a certain sense. Using the ideas that we are about to develop, it is nonetheless possible to study convergence, continuity, compactness, and the other basic concepts of analysis in this more general context. We shall see applications of these new techniques in later sections. ▯

Now we begin to develop the tools of analysis in metric spaces.

DEFINITION 12.5   *Let (X, ρ) be a metric space. A sequence $\{x_j\}$ of elements of X is said to **converge** to a point $\alpha \in X$ if for each $\epsilon > 0$ there is an $N > 0$ such that if $j > N$ then $\rho(x_j, \alpha) < \epsilon$. We call α the **limit** of the sequence $\{x_j\}$. We sometimes write $x_j \to \alpha$.*

Compare this definition of convergence with the corresponding definition for convergence on the real line in Section 3.1. Notice that it is identical, except that the sense in which distance is measured is now more general.

Example 12.6

Let (X, ρ) be the metric space from Example 12.4, consisting of the continuous functions on the unit interval with the indicated metric function ρ. Then $f = \sin x$ is an element of this space, as are the functions

$$f_j = \sum_{\ell=0}^{j} (-1)^\ell \frac{x^{2\ell+1}}{(2\ell+1)!}.$$

Observe that the functions f_j are the partial sums for the Taylor series of $\sin x$. We can check from simple estimates on the error term of Taylor's theorem that the functions f_j converge uniformly to f. Thus, in the language of metric spaces, $f_j \to f$ in the metric space. ⬜

DEFINITION 12.7 *Let (X, ρ) be a metric space. A sequence $\{x_j\}$ of elements of X is said to be **Cauchy** if for each $\epsilon > 0$ there is an $N > 0$ such that if $j, k > N$ then $\rho(x_j, x_k) < \epsilon$.*

Now the Cauchy criterion and convergence are connected in the expected fashion:

PROPOSITION 12.1

Let $\{x_j\}$ be a convergent sequence, with limit α, in the metric space (X, ρ). Then the sequence $\{x_j\}$ is Cauchy.

PROOF Let $\epsilon > 0$. Choose an N so large that if $j > N$ then $\rho(x_j, \alpha) < \epsilon/2$. If $j, k > N$ then

$$\rho(x_j, x_k) \leq \rho(x_j, P) + \rho(P, x_k) < \frac{\epsilon}{2} + \frac{\epsilon}{2} = \epsilon.$$

That completes the proof. ∎

The converse of the proposition is true in the real numbers (with the usual metric), as we proved in Section 3.1. However, it is not true in every metric space. For example, the rationals $\mathbb{Q}$ with the usual metric $\rho(s, t) = |s - t|$ is a metric space; but the sequence

$$3, 3.1, 3.14, 3.141, 3.1415, 3.14159, \ldots,$$

while certainly Cauchy, *does not converge to a rational number*. Thus we are led to a definition:

DEFINITION 12.8 *We say that a metric space (X, ρ) is **complete** if every Cauchy sequence converges to an element of the metric space.*

Thus the real numbers, with the usual metric, form a complete metric space. The rational numbers do not.

Example 12.9

Consider the metric space (X, ρ) from Example 12.4, consisting of the continuous functions on the closed unit interval with the indicated metric function ρ. If $\{g_j\}$ is a Cauchy sequence in this metric space, then each g_j is a continuous function on the unit interval and this sequence of continuous functions is Cauchy in the uniform sense (see Chapter 6). Therefore they converge uniformly to a limit function g that must be continuous. We conclude that the metric space (X, ρ) is complete. ☐

Example 12.10

Consider the metric space (X, ρ) consisting of the polynomials, taken to have domain the interval $[0, 1]$, with the distance function $\rho(f, g) = \sup_{t \in [0,1]} |f(t) - g(t)|$. This metric space is *not* complete. For if h is any continuous function on $[0, 1]$ that is not a polynomial, such as $h(x) = \sin x$, then by the Weierstrass Approximation Theorem there is a sequence $\{p_j\}$ of polynomials that converges uniformly on $[0, 1]$ to h. Thus this sequence $\{p_j\}$ will be Cauchy in the metric space, but it *does not converge to an element of the metric space*. We conclude that the metric space (X, ρ) is not complete. ☐

DEFINITION 12.11   *Let (X, ρ) be a metric space and E a subset of X. A point $P \in E$ is called an **isolated point** of E if there is an $r > 0$ such that $E \cap B(P, r) = \{P\}$. If a point of E is not isolated then it is called **non-isolated**.*

We see that the notion of "isolated" has intuitive appeal: an isolated point is one that is spaced apart — at least distance r — from the other points of the space. A non-isolated point, by contrast, has neighbors that are arbitrarily close.

DEFINITION 12.12   *Let (X, ρ) be a metric space and $f : X \to \mathbb{R}$. If $P \in X$ and $\ell \in \mathbb{R}$ we say that **the limit of f at P is ℓ,** and write*

$$\lim_{x \to P} f(x) = \ell,$$

if for any $\epsilon > 0$ there is a $\delta > 0$ such that if $0 < \rho(x, P) < \delta$ then $|f(x) - \ell| < \epsilon$.

Notice in this definition that we use ρ to measure distance in X — that is, the natural notion of distance with which X comes equipped — but we use absolute values to measure distance in $\mathbb{R}$.

The following lemma will prove useful.

LEMMA 12.2

Let (X, ρ) be a metric space and $P \in X$. Let f be a function from X to $\mathbb{R}$. Then $\lim_{x \to P} f(x) = \ell$ if and only if for every sequence $\{x_j\} \subseteq X$ satisfying $x_j \to P$ it holds that $f(x_j) \to f(P)$.

PROOF   This is straightforward and is treated in the Exercises. ∎

DEFINITION 12.13   *Let (X, ρ) be a metric space and E a subset of X. Suppose that $P \in E$. We say that a function $f : E \to \mathbb{R}$ is continuous at P if*

$$\lim_{x \to P} f(x) = f(P).$$

Example 12.14

Let (X, ρ) be the space of continuous functions on the interval $[0, 1]$ equipped with the supremum metric as in Example 12.4. Define the function $\mathcal{F} : X \to \mathbb{R}$ by the formula

$$\mathcal{F}(f) = \int_0^1 f(t)\, dt.$$

Then $\mathcal{F}$ takes an element of X, namely a continuous function, to a real number, namely its integral over $[0, 1]$. We claim that $\mathcal{F}$ is continuous at every point of X.

For fix a point $f \in X$. If $\{f_j\}$ is a sequence of elements of X converging in the metric space sense to the limit f, then (in the language of classical analysis as in Chapters 6–8) the f_j are continuous functions converging uniformly to the continuous function f on the interval $[0, 1]$. But, by Theorem 9.2, it follows that

$$\int_0^1 f_j(t)\, dt \to \int_0^1 f(t)\, dt.$$

However, this just says that $\mathcal{F}(f_j) \to \mathcal{F}(f)$. Using the lemma, we conclude that

$$\lim_{g \to f} \mathcal{F}(g) = \mathcal{F}(f).$$

Therefore $\mathcal{F}$ is continuous at f.

Since $f \in X$ was chosen arbitrarily, we conclude that the function $\mathcal{F}$ is continuous at every point of X. ☐

In the next section we shall develop some topological properties of metric spaces.

12.2   Topology in a Metric Space

Fix a metric space (X, ρ). An *open ball* in the metric space is a set of the form

$$B(P, r) \equiv \{x \in X : \rho(x, P) < r\},$$

where $P \in X$ and $r > 0$. A set $U \subseteq X$ is called *open* if for each $u \in U$ there is an $r > 0$ such that $B(u, r) \subseteq U$.

We define a *closed ball* in the metric space (X, ρ) to be $\overline{B}(P, r) \equiv \{x \in X : \rho(x, P) \leq r\}$. A set $E \subseteq X$ is called *closed* if its complement in X is open.

Example 12.15

Consider the set of real numbers $\mathbb{R}$ equipped with the metric $\rho(s, t) = 1$ if $s \neq t$ and $\rho(s, t) = 0$ otherwise. Then each singleton $U = \{x\}$ is an open set. For let P be a point of U. Then $P = x$ and the ball $B(P, 1/2)$ lies in U.

However, each singleton is also closed. For the complement of the singleton $U = \{x\}$ is the set $S = \mathbb{R} \setminus \{x\}$. If $s \in S$ then $B(s, 1/2) \subseteq S$ as in the preceding paragraph.   ⬚

Example 12.16

Let (X, ρ) be the metric space of continuous functions on the interval $[0, 1]$ equipped with the metric $\rho(f, g) = \sup_{x \in [0,1]} |f(x) - g(x)|$. Define

$$U = \{f \in X : f(1/2) > 5\}.$$

Then U is an open set in the metric space. To verify this, fix an element $f \in U$. Let $\epsilon = f(1/2) - 5 > 0$. We claim that the metric ball $B(f, \epsilon)$ lies in U. For let $g \in B(f, \epsilon)$. Then

$$g(1/2) \geq f(1/2) - |f(1/2) - g(1/2)|$$
$$\geq f(1/2) - \rho(f, g)$$
$$> f(1/2) - \epsilon$$
$$= 5.$$

It follows that $g \in U$. Since $g \in B(f, \epsilon)$ was chosen arbitrarily, we may conclude that $B(f, \epsilon) \subseteq U$. But this says that U is open.

We may also conclude from this calculation that

$$^cU = \{f \in X : f(1/2) \leq 5\}$$

is closed.   ⬚

DEFINITION 12.17 *Let (X, ρ) be a metric space and $S \subseteq X$. A point $x \in X$ is called an **accumulation point** of S if every $B(x, r)$ contains infinitely many elements of S.*

PROPOSITION 12.3
Let (X, ρ) be a metric space. A set $S \subseteq X$ is closed if and only if every accumulation point of S lies in S.

PROOF The proof is similar to ideas discussed in Section 5.1 and we leave it to the Exercises. ∎

DEFINITION 12.18 *Let (X, ρ) be a metric space. A subset $S \subseteq X$ is said to be **bounded** if S lies in some ball $B(P, r)$.*

DEFINITION 12.19 *Let (X, ρ) be a metric space. A set $S \subseteq X$ is said to be **compact** if every sequence in S has a subsequence that converges **to an element of S**.*

Example 12.20
In Chapter 5 we learned that in the real number system compact sets are closed and bounded, and conversely. Such is not the case in general metric spaces.

As an example, consider the metric space (X, ρ) consisting of all continuous functions on the interval $[0, 1]$ with the supremum metric as in previous examples. Let

$$S = \{f_j(x) = x^j : j = 1, 2, \ldots\}.$$

This set is bounded since it lies in the ball $B(\mathbf{0}, 2)$ (here $\mathbf{0}$ denotes the identically zero function). We claim that S contains no Cauchy sequences. This follows (see the discussion of uniform convergence in Chapter 9) because, no matter how large N is, if $k > j > N$ then we may write

$$|f_j(x) - f_k(x)| = |x^j| \, |(x^{k-j} - 1)|.$$

Fix j. If x is sufficiently near to 1 then $|x^j| > 3/4$. But then we may pick k so large that $|x^{k-j}| < 1/4$. Thus

$$|f_k(x) - f_j(x)| > 9/16.$$

So there is no Cauchy subsequence. We may conclude (for vacuous reasons) that S is closed.

But S is not compact. For, as just noted, the sequence $\{f_j\}$ consists of infinitely many distinct elements of S that do not have a convergent subsequence (indeed not even a Cauchy subsequence). ☐

In spite of the last example, half of the Heine–Borel Theorem is true:

PROPOSITION 12.4
Let (X, ρ) be a metric space and S a subset of X. If S is compact then S is closed and bounded.

PROOF   Let $\{s_j\}$ be a Cauchy sequence in S. By compactness, this sequence must contain a subsequence converging to some limit P. But since the full sequence is Cauchy, the full sequence must converge to P (Exercise). Thus S is closed.

If S is not bounded, we derive a contradiction as follows. Fix a point $P_1 \in S$. Since S is not bounded, we may find a point P_2 that has distance at least 1 from P_1. Since S is unbounded, we may find a point P_3 of S that is distance at least 2 from both P_1 and P_2. Continuing in this fashion, we select $P_j \in S$ which is distance at least j from $P_1, P_2, \ldots P_{j-1}$. Such a sequence $\{P_j\}$ can have no Cauchy subsequence, contradicting compactness. Therefore S is bounded.   ∎

DEFINITION 12.21   *Let S be a subset of a metric space (X, ρ). A collection of open sets $\{\mathcal{O}_\alpha\}_{\alpha \in A}$ (each $\mathcal{O}_\alpha$ is an open set in X) is called an **open covering** of S if*

$$\cup_{\alpha \in A} \mathcal{O}_\alpha \supseteq S.$$

DEFINITION 12.22   *If $\mathcal{C}$ is an open covering of a set S and if $\mathcal{D}$ is another open covering of S such that each element of $\mathcal{D}$ is also an element of $\mathcal{C}$ then we call $\mathcal{D}$ a **subcovering** of $\mathcal{C}$.*
*We call $\mathcal{D}$ a **finite subcovering** if $\mathcal{D}$ has just finitely many elements.*

THEOREM 12.5
A set subset S of a metric space (X, ρ) is compact if and only if every open covering $\mathcal{C} = \{\mathcal{O}_\alpha\}_{\alpha \in A}$ of S has a finite subcovering.

PROOF   The forward direction is beyond the scope of this book and we shall not discuss it.

The proof of the reverse direction is similar to the proof in Section 5.3 (Theorem 5.14). We leave the details for the Exercises.   ∎

PROPOSITION 12.6
Let S be a compact subset of a metric space (X, ρ). If E is a closed subset of S then E is compact.

PROOF   Let $\mathcal{C}$ be an open covering of E. The set $U = X \setminus E$ is open and the covering $\mathcal{C}'$ consisting of all the open sets in $\mathcal{C}$ together with the open set U

covers S. Since S is compact we may find a finite subcovering

$$O_1, O_2, \ldots O_k$$

that covers S. If one of these sets is U then discard it. The remaining $k - 1$ open sets cover E. ∎

The Exercises will ask you to find an alternate proof of this last fact.

12.3   The Baire Category Theorem

Let (X, ρ) be a metric space and $S \subseteq X$ a subset. A set $E \subseteq X$ is said to be *dense* in S if every element of S is the limit of some sequence of elements of E.

Example 12.23

The set of rational numbers $\mathbb{Q}$ is dense in any subset of the reals $\mathbb{R}$ equipped with the usual metric.   ⬚

Example 12.24

Let (X, ρ) be the metric space of continuous functions on the interval $[0, 1]$ equipped with the supremum metric as usual. Let $E \subseteq X$ be the polynomial functions. Then the Weierstrass Approximation Theorem tells us that E is dense in X.   ⬚

Example 12.25

Consider the real numbers $\mathbb{R}$ with the metric $\rho(s, t) = 1$ if $s \neq t$ and $\rho(s, t) = 0$ otherwise. Then no proper subset of $\mathbb{R}$ is dense in $\mathbb{R}$. To see this, notice that if E were dense and were not all of $\mathbb{R}$ and if $P \in \mathbb{R} \setminus E$ then $\rho(P, e) > 1/2$ for all $e \in E$. So elements of E do not get close to P. Thus E is not dense in $\mathbb{R}$.   ⬚

DEFINITION 12.26   *If (X, ρ) is a metric space and $E \subseteq X$ then the **closure** of E is defined to be the union of E with the set of its accumulation points.*

Example 12.27

Let (X, ρ) be the set of real numbers with the usual metric, and set $E = \mathbb{Q} \cap (-2, 2)$. Then the closure of E is $[-2, 2]$.

Let (Y, σ) be the continuous functions on $[0, 1]$ equipped with the supremum metric as in Example 12.4. Take $E \subseteq Y$ to be the polynomials. Then the closure of E is Y.   ☐

We note in passing that if $B(P, r)$ is a ball in a metric space (X, ρ), then $\overline{B}(P, r)$ will contain, but need not be equal to, the closure of $B(P, r)$ (for which see Exercise 6).

DEFINITION 12.28   *Let (X, ρ) be a metric space. We say that $E \subseteq X$ is* **nowhere dense** *in X if the closure of E contains no ball $B(x, r)$ for any $x \in X, r > 0$.*

Example 12.29

Let us consider the integers $\mathbb{Z}$ as a subset of the metric space $\mathbb{R}$ equipped with the standard metric. Then the closure of $\mathbb{Z}$ is $\mathbb{Z}$ itself. And, of course, $\mathbb{Z}$ contains no metric balls. Therefore $\mathbb{Z}$ is nowhere dense in $\mathbb{R}$.   ☐

Example 12.30

Consider the metric space X of all continuous functions on the unit interval $[0, 1]$, equipped with the usual supremum metric. Fix $k > 0$ and consider

$$E \equiv \{p(x) : p \text{ is a polynomial of degree not exceeding } k\}.$$

Then the closure of E is E itself (that is, the limit of a sequence of polynomials of degree not exceeding k is still a polynomial of degree not exceeding k — details are requested of you in the Exercises). And E contains no metric balls. For if $p \in E$ and $r > 0$ then $p(x) + (r/2) \cdot x^{k+1} \in B(p, r)$ but $p(x) + (r/2) \cdot x^{k+1} \notin E$.

We recall, as noted in Example 12.24 above, that the set of *all* polynomials is dense in X, but if we restrict attention to polynomials of degree not exceeding a fixed number k then the resulting set is nowhere dense.   ☐

THEOREM 12.7   THE BAIRE CATEGORY THEOREM

Let (X, ρ) be a complete metric space. Then X cannot be written as the union of countably many nowhere-dense sets.

PROOF This proof is quite similar to the proof that we presented in Chapter 5 that a perfect set must be uncountable. You may wish to review that proof at this time.

Seeking a contradiction, suppose that X may be written as a countable union of nowhere-dense sets $Y_1, Y_2, \ldots$. Choose a point $x_1 \in {}^c \bar{Y}_1$. Since Y_1 is nowhere dense, we may select an $r_1 > 0$ such that $\bar{B}_1 \equiv \bar{B}(x_1, r_1)$ satisfies $\bar{B}_1 \cap \bar{Y}_1 = \emptyset$. Assume without loss of generality that $r_1 < 1$.

Next, since Y_2 is nowhere dense, we may choose $x_2 \in \bar{B}_1 \cap {}^c \bar{Y}_2$ and an $r_2 > 0$ such that $\bar{B}_2 = \bar{B}(x_2, r_2) \subseteq \bar{B}_1 \cap {}^c \bar{Y}_2$. Shrinking B_2 if necessary, we may assume that $r_2 < \frac{1}{2} r_1$. Continuing in this fashion, we select at the jth step a point $x_j \in \bar{B}_{j-1} \cap {}^c \bar{Y}_j$ and a number $r_j > 0$ such that $r_j < \frac{1}{2} r_{j-1}$ and $\bar{B}_j = \bar{B}(x_j, r_j) \subseteq \bar{B}_{j-1} \cap {}^c \bar{Y}_j$.

Now the sequence $\{x_j\}$ is Cauchy since all the terms x_j for $j > N$ are contained in a ball of radius $r_N < 2^{-N+1}$, hence are not more than distance 2^{-N+1} apart. Since (X, ρ) is a complete metric space, we conclude that the sequence converges to a limit point P. Moreover, by construction, $P \in \bar{B}_j$ for every j, hence is in the complement of *every* $\bar{Y}_j$. Thus $\cup_j \bar{Y}_j \neq X$. That is a contradiction, and the proof is complete. ∎

Before we apply the Baire Category Theorem, let us formulate some re-statements, or corollaries, of the theorem which follow immediately from the definitions.

COROLLARY 12.8
Let (X, ρ) be a complete metric space. Let $Y_1, Y_2, \ldots$ be countably many closed subsets of X, each of which contains no nontrivial open ball. Then $\bigcup_j Y_j$ also has the property that it contains no nontrivial open ball.

COROLLARY 12.9
Let (X, ρ) be a complete metric space. Let $O_1, O_2, \ldots$ be countably many dense open subsets of X. Then $\bigcap_j O_j$ is dense in X.

Note that the result of the second corollary follows from the first corollary by complementation. The set $\bigcap_j O_j$, while dense, need not be open.

Example 12.31
The metric space $\mathbb{R}$, equipped with the standard Euclidean metric, cannot be written as a countable union of nowhere-dense sets. ▯

By contrast, $\mathbb{Q}$ *can* be written as the union of the singletons $\{q_j\}$ where the q_j represent an enumeration of the rationals. However, $\mathbb{Q}$ is not complete.

Example 12.32

Baire's theorem contains the fact that a perfect set of real numbers must be uncountable. For if P is perfect and countable we may write $P = \{p_1, p_2, \ldots\}$. Therefore

$$P = \bigcup_{j=1}^{\infty} \{p_j\}.$$

But each of the singletons $\{p_j\}$ is a nowhere-dense set in the metric space P. And P is complete. (You should verify both these assertions for yourself.) This contradicts the Category Theorem. So P cannot be countable.   ▯

A set that can be written as a countable union of nowhere-dense sets is said to be of *first category*. If a set is not of first category, then it is said to be of *second category*. The Baire Category Theorem says that a complete metric space must be of second category. We should think of a set of first category as being "thin" and a set of second category as being "fat" or "robust." (This is one of many ways that we have in mathematics of distinguishing "fat" sets. Countability and uncountability is another. Lebesgue's measure theory, not covered in this book, is a third.)

One of the most striking applications of the Baire Category Theorem is the following result to the effect that "most" continuous functions are nowhere differentiable. This explodes the myth that most of us learn in calculus that a typical function is differentiable at all points except perhaps at a discrete set of bad points.

THEOREM 12.10

Let (X, ρ) be the metric space of continuous functions on the unit interval $[0, 1]$ equipped with the metric

$$\rho(f, g) = \sup_{x \in [0,1]} |f(x) - g(x)|.$$

Define a subset of E of X as follows: $f \in E$ if there exists one point at which f is differentiable. Then E is of first category in the complete metric space (X, ρ).

PROOF   For each positive integer m, n we let

$$A_{m,n} = \{f \in X : \exists x \in [0, 1] \text{ such that } |f(x) - f(t)| \leq n|x - t|$$

$$\forall\, t \in [0, 1] \text{ that satisfy } |x - t| \leq 1/m\}.$$

Fix m and n. We claim that $A_{m,n}$ is nowhere dense in X. In fact, if $f \in A_{m,n}$ set

$$K_f = \max_{x \in [0,1]} \left| \frac{f(x \pm 1/m) - f(x)}{1/m} \right|.$$

Let $h(x)$ be a continuous piecewise linear function, bounded by 1, consisting of linear pieces having slope $3K_f$. Then for every $\epsilon > 0$ it holds that $f + \epsilon \cdot h$ has metric distance less than ϵ from f and is not a member of $A_{m,n}$. This proves that $A_{m,n}$ is nowhere dense.

We conclude from Baire's theorem that $\cup_{m,n} A_{m,n}$ is nowhere dense in X. Therefore $S = X \setminus \cup_{m,n} A_{m,n}$ is of second category. But if $f \in S$ then for every $x \in [0, 1]$ and every $n > 0$ there are points t arbitrarily close to x (that is, at distance $\leq 1/m$ from x) such that

$$\left| \frac{f(t) - f(x)}{t - x} \right| > n.$$

It follows that f is differentiable at no $x \in [0, 1]$. That proves the assertion. ∎

12.4   The Ascoli–Arzela Theorem

Let $\mathcal{F} = \{f_\alpha\}_{\alpha \in A}$ be a family, not necessarily countable, of functions on a metric space (X, ρ). We say that the family $\mathcal{F}$ is *equicontinuous* on X if for every $\epsilon > 0$ there is a $\delta > 0$ such that when $\rho(s, t) < \delta$ then $|f_\alpha(s) - f_\alpha(t)| < \epsilon$. Notice that equicontinuity mandates not only uniform continuity of each f_α but also that the uniformity occur simultaneously, and at the same rate, for all the f_α.

Example 12.33
Let (X, ρ) be the unit interval $[0, 1]$ with the usual Euclidean metric. Let $\mathcal{F}$ consist of all functions f on X that satisfy the Lipschitz condition

$$|f(s) - f(t)| \leq 2 \cdot |s - t|$$

for all s, t. Then $\mathcal{F}$ is an equicontinuous family of functions. For if $\epsilon > 0$, then we may take $\delta = \epsilon/2$. Then if $|s - t| < \delta$ and $f \in \mathcal{F}$, we have

$$|f(s) - f(t)| \leq 2 \cdot |s - t| < 2 \cdot \delta = \epsilon.$$

Observe, for instance, that the Mean Value Theorem tells us that $\sin x, \cos x$, $2x, x^2$ are elements of $\mathcal{F}$. ⬜

If $\mathcal{F}$ is a family of functions on X we call $\mathcal{F}$ *equibounded* if there is a number $M > 0$ such that

$$|f(x)| \leq M$$

for all $x \in X$ and all $f \in \mathcal{F}$. For example, the functions $f_j(x) = \sin jx$ on $[0, 1]$ form an equibounded family.

One of the cornerstones of classical analysis is the following result of Ascoli and Arzela:

THEOREM 12.11   THE ASCOLI–ARZELA THEOREM

Let (Y, σ) be a metric space and assume that Y is compact. Let $\mathcal{F}$ be an equibounded, equicontinuous family of functions on Y. Then there is a sequence $\{f_j\} \subseteq \mathcal{F}$ that converges uniformly to a continuous function on Y.

Before we prove this theorem, let us comment on it. Let (X, ρ) be the metric space consisting of the continuous functions on the unit interval $[0, 1]$ equipped with the usual supremum norm. Let $\mathcal{F}$ be an equicontinuous, equibounded family of functions on $[0, 1]$. Then the theorem says that $\mathcal{F}$ is a compact set in this metric space. For any infinite subset of $\mathcal{F}$ is guaranteed to have a convergent subsequence. As a result, we may interpret the Ascoli–Arzela theorem as identifying certain compact collections of continuous functions.

PROOF OF THE ASCOLI–ARZELA THEOREM   We divide the proof into a sequence of lemmas. The hypotheses of the theorem will stand throughout these arguments.

LEMMA 12.12

*Let $\eta > 0$. There exist finitely many points $y_1, y_2, \ldots y_k \in Y$ such that every ball $B(s, \eta) \subseteq Y$ contains one of the y_j. We call $y_1, \ldots, y_k$ an η-**net** for Y.*

PROOF   Consider the collection of balls $\{B(y, \eta/2) : y \in Y\}$. This is an open covering of Y, hence by compactness has a finite subcovering $B(y_1, \eta/2), \ldots,$ $B(y_k, \eta/2)$. The centers $y_1, \ldots, y_k$ are the points we seek. For if $B(s, \eta)$ is *any* ball in Y then its center s must be contained in some ball $B(y_j, \eta/2)$. But then $B(y_j, \eta/2) \subseteq B(s, \eta)$; hence, in particular, $y_j \in B(s, \eta)$.   ∎

LEMMA 12.13

Let $\epsilon > 0$. There is an $\eta > 0$, a corresponding η-net $y_1, \ldots y_k$, and a sequence $\{f_m\} \subseteq \mathcal{F}$ such that

- *The sequence $\{f_m(y_\ell)\}_{m=1}^{\infty}$ converges for each y_ℓ.*
- *For any $y \in Y$ the sequence $\{f_m(y)\}_{j=1}^{\infty}$ is contained in an interval in the real line of length at most ϵ.*

PROOF   By equicontinuity there is an $\eta > 0$ such that if $\rho(s, t) < \eta$ then $|f(s) - f(t)| < \epsilon/3$ for every $f \in \mathcal{F}$. Let $y_1, \ldots, y_k$ be an η-net. Since the family $\mathcal{F}$ is equibounded, the set of numbers $\{f(y_1) : f \in \mathcal{F}\}$ is bounded. Thus there is a subsequence f_j such that $\{f_j(y_1)\}$ converges. But then, by similar reasoning, we may choose a subsequence f_{j_k} such that $\{f_{j_k}(y_2)\}$ converges.

Continuing in this fashion, we may find a sequence, which we call $\{f_m\}$, which converges at each point y_ℓ. The first assertion is proved. Discarding finitely many of the f_m's, we may suppose that for every m, n and every j it holds that $|f_m(y_j) - f_n(y_j)| < \epsilon/3$.

Now if y is *any* point of Y then there is an element y_t of the η-net such that $\rho(y, y_t) < \eta$. But then, for any m, n, we have

$$|f_m(y) - f_n(y)| \leq |f_m(y) - f_m(y_t)| + |f_m(y_t) - f_n(y_t)| + |f_n(y_t) - f_n(y)|$$
$$< \frac{\epsilon}{3} + \frac{\epsilon}{3} + \frac{\epsilon}{3}$$
$$= \epsilon.$$

That proves the second assertion. ∎

PROOF OF THE THEOREM   With $\epsilon = 2^{-1}$, apply Lemma 12.13 to obtain a sequence f_m. Apply Lemma 12.13 again, with $\epsilon = 2^{-2}$ and the role of $\mathcal{F}$ being played by the sequence $\{f_m\}$. This yields a new sequence $\{f_{m_r}\}$. Apply Lemma 12.13 once again with $\epsilon = 2^{-3}$ and the role of $\mathcal{F}$ being played by the second sequence $\{f_{m_r}\}$. Keep going to produce a countable list of sequences.

Now produce the final sequence by selecting the first element of the first sequence, the second element of the second sequence, the third element of the third sequence, and so forth. This sequence, which we call $\{f_w\}$, will satisfy the conclusion of the theorem.

For if $\epsilon > 0$ then there is a j such that $2^{-j} < \epsilon$. After j terms, the sequence $\{f_w\}$ is a subsequence of the jth sequence constructed above. Hence at every $y \in Y$, all the terms $f_w(y), w > j$, lie in an interval of length ϵ. But that just verifies convergence at the point y. Note moreover that the choice of j in this last argument was independent of $y \in Y$. That shows that the convergence is uniform. The proof is complete. ∎

The technique used to construct the sequence $\{f_w\}$ is frequently called a *diagonalization argument*. It is a frequently used device in mathematical analysis.

Exercises

12.1 Let (X, ρ) be a metric space. Prove that the function
$$\sigma(s, t) = \frac{\rho(s, t)}{1 + \rho(s, t)}$$
is also a metric on X and that the open sets defined by the metric ρ are the same as the open sets defined by σ. Finally, prove that $\sigma(s, t) < 1$ for all $s, t \in X$.

12.2 Let (X, ρ) be a metric space, and $E \subseteq X$. Define the *interior* of E to be those points $e \in E$ such that there exists an $r > 0$ with $B(e, r) \subseteq E$. Prove that the interior of any set is open. Give an example of a set in a metric space that is not equal to its interior.

12.3 Let (X, ρ) be a metric space and E a subset of X. Define the *boundary* of E to be those elements $x \in X$ with the property that every ball $B(x, r)$ contains both points of E and points of ^cE. Prove that the boundary of E must be closed. Prove that the interior of E is disjoint from the boundary of E.

12.4 Let (X, ρ) be a metric space. Prove that the closure of any set in X is closed. Prove that the closure of any E equals the union of the interior and the boundary.

12.5 Let (X, ρ) be a metric space. Let $K_1 \supseteq K_2 \ldots$ be a nested family of countably many nonempty compact sets. Prove that $\cap_j K_j$ is a nonempty set.

12.6 Give an example of a metric space (X, ρ), a point $P \in X$, and a positive number r such that $\overline{B}(P, r)$ is *not* the closure of the ball $B(P, r)$.

12.7 Let (X, ρ) be the collection of continuous functions on the interval $[0, 1]$ equipped with the usual supremum metric. Let $E_j = \{p(x) : p \text{ is a polynomial of degree not exceeding } k\}$. Then, as noted in the text, each E_j is nowhere dense in X. Yet $\cup_j E_j$ *is* dense in X. Prove these assertions and explain why they do not contradict Baire's Theorem.

12.8 Assume f_j is a sequence of continuous, real-valued functions on $\mathbb{R}$ with the property that $\{f_j(x)\}$ is unbounded whenever $x \in \mathbb{Q}$. Use the Category Theorem to prove that it cannot then be true that whenever t is irrational then the sequence $\{f_j(t)\}$ is bounded.

12.9 Consider the space X of all integrable functions on the interval $[0, 1]$. Define a metric, for $f, g \in X$, by the equation

$$\rho(f, g) = \int_0^1 |f(x) - g(x)| \, dx.$$

Prove that this is indeed a metric. The set S of continuous functions lies in X; we usually equip S with the supremum metric. How does the supremum metric compare with this new metric? Show that S is dense in X.

12.10 Let (X, ρ) be a metric space. Let $f : X \to \mathbb{R}$ be a function. Prove that f is continuous if and only if $f^{-1}(U)$ is open whenever $U \subseteq \mathbb{R}$ is open.

12.11 Let (X, ρ) be a compact metric space. Prove that X has a countable dense subset. [We call such a space *separable*.]

12.12 Let K be a compact subset of a metric space (X, ρ). Let $P \in X$ not lie in K. Prove that there is an element $k \in K$ such that

$$\rho(k, P) = \inf_{x \in K} \rho(x, P).$$

12.13 Consider the metric space $\mathbb{Q}$ equipped with the Euclidean metric. Give an example of a set in this metric space that is closed and bounded but is not compact.

12.14 Consider the metric space $\mathbb{Q}$ equipped with the Euclidean metric. Describe all the open sets in this metric space.

12.15 A certain metric space has the property that the only open sets are singletons. What can you conclude about this metric space?

12.16 In $\mathbb{R}$, if I is an open interval then every element of I is a limit point of I. Is the analogous statement true in an arbitrary metric space, with "interval" replaced by "ball"?

12.17 The Bolzano–Weierstrass Theorem tells us that, in $\mathbb{R}^1$, a bounded infinite set must have a limit point. Show by example that the analogous statement is false in an arbitrary metric space.

12.18 Let (X, ρ) and (Y, σ) be metric spaces. Describe a method for equipping the set $X \times Y$ with a metric manufactured from ρ and σ.

12.19 Refer to Exercises 2 and 3 for terminology. Let E be a subset of a metric space. Is the interior of E equal to the interior of the closure of E? Is the closure of the interior of E equal to the closure of E itself?

12.20 Let X be the collection of all continuously differentiable functions on the interval $[0, 1]$. If $f, g \in X$ then define

$$\rho(f, g) = \sup_{x \in [0,1]} |f'(x) - g'(x)|.$$

Is ρ a metric? Why or why not?

12.21 Let (X, ρ) be a metric space. Call a subset E of X *connected* if there do not exist open sets U and V in X such that $U \cap E$ and $V \cap E$ are nonempty, disjoint, and $(U \cap E) \cup (V \cap E) = E$. Is the closure of a connected set connected? Is the product of two connected sets connected? Is the interior of a connected set connected?

12.22 Refer to Exercise 21 for terminology. Give exact conditions that will guarantee that the union of two connected sets is connected.

12.23 Consider a collection $\mathcal{F}$ of differentiable functions on the interval $[a, b]$ that satisfy the conditions $|f(x)| \leq K$ and $|f'(x)| \leq C$ for all $x \in [a, b], f \in \mathcal{F}$. Demonstrate that the Ascoli–Arzela theorem applies to $\mathcal{F}$, and describe the resulting conclusion.

12.24 Even if we did not know the transcendental functions $\sin x, \cos x, \ln x, e^x$, etc. explicitly, the Baire Category Theorem demonstrates that transcendental functions must exist. Explain why this assertion is true.

12.25 Refer to Exercise 9 for definitions. On this metric space, define

$$T : X \to \mathbb{R}$$

by the formula

$$T(f) = \int_0^1 f(x)\, dx.$$

Is T a continuous function from X to $\mathbb{R}$?

12.26 Let (X, ρ) be the metric space of continuously differentiable functions on the interval $[0, 1]$ equipped with the metric

$$\rho(f, g) = \sup_{x \in [0,1]} |f(x) - g(x)|.$$

Consider the function

$$T(f) = f'(1/2).$$

Is T continuous? Is there some metric with which we can equip X that will make T continuous?

12.27 Prove Lemma 12.2.

12.28 Prove Proposition 12.3.

12.29 Let (X, ρ) be a metric space and let $\{x_j\}$ be a Cauchy sequence in X. If a subsequence $\{x_{j_k}\}$ converges to a point $P \in X$, then prove that the full sequence $\{x_j\}$ converges to P.

12.30 Prove the converse direction of Theorem 12.5.

12.31 Give a proof of Proposition 12.6 that uses the sequential definition of compactness.

12.32 Let $\{p_j(x)\}$ be a sequence of polynomial functions on $\mathbb{R}^3$, each of degree not exceeding k. Assume that this sequence converges pointwise to a limit function f. Prove that f is a polynomial of degree not exceeding k. Can you weaken the hypothesis on the p_j's so that they need not be polynomials?

12.33 Let (X, ρ) be any metric space. Consider the space $\hat{X}$ of all Cauchy sequences of elements of X, subject to the equivalence relation that $\{x_j\}$ and $\{y_j\}$ are equivalent if $\rho(x_j, y_j) \to 0$ as $j \to \infty$. Explain why, in a natural way, this space of equivalence class of Cauchy sequences may be thought of as *the completion* of X; that is, explain in what sense $\hat{X} \supseteq X$ and $\hat{X}$ is complete. Prove that $\hat{X}$ is minimal in a certain sense. Prove that if X is already complete then this space of equivalence classes can be identified in a natural way with X.

Bibliography

[BOA] R. P. Boas, *A Primer of Real Functions*. Carus Mathematical Monograph No. 13, John Wiley and Sons, Inc., NY, 1960.

[BUC] R. C. Buck. *Advanced Calculus*. 2d ed., McGraw-Hill Book Company, NY, 1965.

[HOF] K. Hoffman. *Analysis in Euclidean Space*. Prentice-Hall, Inc., Englewood Cliffs, NJ, 1975.

[NIV] I. Niven. *Irrational Numbers*. Carus Mathematical Monograph No. 11, John Wiley and Sons, Inc., NY, 1956.

[RUD] W. Rudin. *Principles of Mathematical Analysis*. 3d ed., McGraw-Hill Book Company, NY, 1976.

[STR] K. Stromberg. *An Introduction to Classical Real Analysis*. Wadsworth Publishing, Inc., Belmont, CA, 1981.

Index